Spacecraft Anomalies and Failures Workshop 2023

NASA Introductory Comments

Joseph I. Minow, PhD

NIMBLE BOOKS LLC: THE AI LAB FOR BOOK-LOVERS

~ FRED ZIMMERMAN, EDITOR ~

Humans and AI making books richer, more diverse, and more surprising.

Publishing Information

(c) 2024 Nimble Books LLC
ISBN: 978-1-60888-288-5

AI-generated Keyword Phrases

spacecraft anomalies; spacecraft failures; space weather events; on-orbit breakups; case studies; challenges of space operations; risks of space operations; hazards of the space environment; understanding the space environment; mitigating space hazards; evaluating the space environment; effects on space systems.

Publisher's Notes

For anyone interested in space power, safety is, or should be, a paramount concern, both because of the inherent value of human life and material, and because a loss of confidence could have disastrous consequences. For these reasons, it is important for students of the topic to have an empirically grounded understanding of risk. In a previous era, NASA, relying on "engineering judgment", initially estimated the risk of loss of the space shuttle at between 1 in 1000 and 1 in 100,000; the actual rate was 2 in 133, both with complete loss of vessel and crew. For this reason, this 2023 presentation from a joint NASA/NRO conference on recent spacecraft mission anomalies should be essential reading for all practitioners. Readers should expect to:

- Gain valuable insights into recent spacecraft anomalies and failures, providing a deeper understanding of the challenges and risks associated with space operations.
- Understand the importance of safety measures and continuous improvement in space technology, particularly in the current era of increasing space exploration and commercial space activities.
- Recognize the significance of space situational awareness, space weather monitoring, and robust space debris mitigation strategies in addressing the growing number of satellites and debris in orbit.

This annotated edition illustrates the capabilities of the AI Lab for Book-Lovers to add context and ease-of-use to manuscripts. It includes several types of abstracts, building from simplest to more complex: TLDR (one word), ELI5, TLDR (vanilla), Scientific Style, and Action Items; essays to increase viewpoint diversity, such as Grounds for Dissent, Red Team Critique, and MAGA Perspective; and Notable Passages and Nutshell Summaries for each page.

ANNOTATIONS

ABSTRACTS

TL;DR (ONE WORD)

Hazards.

EXPLAIN IT TO ME LIKE I'M FIVE YEARS OLD

This document is about things that can go wrong with spacecraft and what causes them. It also talks about bad weather in space and when things break apart in space. There are examples of when these things happened and how it's hard to do space missions because of these problems. The document tells us why it's important to know about the dangers in space and how we can try to make things safer. It also talks about tools and resources that help us understand what happens in space. If you need help or

TL;DR (VANILLA)

This document discusses spacecraft anomalies, failures, space weather events, and on-orbit breakups. It includes case studies and highlights the challenges and risks of space operations. It emphasizes the importance of understanding and mitigating hazards for successful missions. Tools and resources for evaluating the space environment are mentioned, along with contact information for NASA personnel involved in this topic.

SCIENTIFIC STYLE

This document presents information on spacecraft anomalies and failures, as well as space weather events and on-orbit breakups. It includes case studies of specific incidents and discusses the challenges and risks associated with space operations. The document emphasizes the need to understand and mitigate the hazards of the space environment for successful space missions. Various tools and resources for evaluating the space environment and its effects on space systems are mentioned. The document also provides contact information for NASA personnel involved in the Spacecraft Anomalies and Failures Workshop.

ACTION ITEMS

Review the case studies and learn from past incidents to prevent similar failures in future space missions.

Familiarize yourself with the tools and resources mentioned to evaluate the space environment and its effects on space systems.

Contact NASA personnel involved in the Spacecraft Anomalies and Failures Workshop for further information or assistance.

VIEWPOINTS

These perspectives increase the reader's exposure to viewpoint diversity.

GROUNDS FOR DISSENT

A member of the organization responsible for this document might have principled, substantive reasons to dissent from this report due to the following factors:

Lack of Inclusion and Bias: The dissenting member may argue that the document fails to include a comprehensive analysis of all relevant spacecraft anomalies, failures, and space weather events. They might contend that the report has an inherent bias towards certain incidents or focuses on specific technologies, neglecting others. This lack of inclusion may lead to an incomplete understanding of the risks involved in space operations.

Inadequate Analysis: The dissenting member could claim that the document's case studies do not provide sufficient depth or detail to adequately understand the challenges and risks associated with space operations. They might argue that a more rigorous analysis is required to identify patterns, root causes, and potential solutions for preventing future incidents. Without this level of analysis, they may assert that the report does not fulfill its purpose effectively.

Insufficient Mitigation Strategies: Another reason for dissent could be the perceived insufficiency of the document's discussion on mitigating hazards in space environments. The dissenting member may argue that while recognizing and understanding risks is important, there is a lack of emphasis on concrete strategies and measures to address these risks effectively. This deficiency could lead to a disconnect between identifying hazards and implementing practical solutions.

Limited Evaluation Tools: The dissenter might contend that the document falls short in discussing the availability and suitability of tools and resources for evaluating the space environment's effects on space systems. They could argue that a more comprehensive overview of evaluation methods is necessary for providing organizations with accurate assessments and data-driven decision-making processes.

Lack of Stakeholder Engagement: Lastly, the dissenting member could highlight concerns about insufficient stakeholder engagement throughout the report. They may argue that key perspectives from industry experts, astronauts, engineers, or other relevant stakeholders were overlooked during the information gathering process. This oversight could lead to a limited understanding of the challenges faced by those directly involved in space operations and hinder the report's overall credibility.

In summary, a member of the organization responsible for this document might dissent due to concerns about bias, inadequate analysis, insufficient mitigation strategies, limited evaluation tools, and a lack of stakeholder engagement. These substantive reasons illustrate potential weaknesses that could undermine the document's effectiveness and reliability in informing decision-making processes within the organization.

RED TEAM CRITIQUE

Upon reviewing the document on spacecraft anomalies and failures, as well as space weather events and on-orbit breakups, several areas of concern have been identified.

Firstly, while the document provides a comprehensive overview of various incidents and their impacts on space operations, it lacks sufficient analysis and insights into the root causes of these anomalies. While case studies are mentioned, there is a need for more detailed examination and evaluation of each incident's underlying factors to truly understand how future occurrences can be prevented. Additionally, there is limited discussion regarding lessons learned from previous incidents which could aid in improving space mission success rates.

Furthermore, although the document mentions challenges and risks associated with space operations, it fails to provide a thorough assessment of these hazards. Understanding and mitigating risks is crucial for ensuring the safety of both human astronauts and valuable equipment. It would be beneficial to include more specific information regarding potential risks unique to different types of missions or environments.

Moreover, while tools and resources for evaluating the space environment are briefly mentioned within the document, they are not explored in great detail nor their effectiveness discussed. The addition of case studies that demonstrate how these tools have been utilized successfully could enhance readers' understanding.

Additionally, providing contact information for NASA personnel involved in spacecraft anomalies should be approached with caution due to potential security concerns or excessive inquiries that may arise from widespread availability. A more selective approach towards disclosing such sensitive details would help protect vital operations without compromising transparency or access when necessary.

Lastly, it is suggested that this document could benefit from incorporating input from experts outside regular stakeholders involved in spacecraft design or operation processes. Incorporating external perspectives through interviews or collaboration with researchers specialized in aerospace engineering could provide industry-wide insights towards preventing future incidents effectively.

In conclusion, while this document serves as an informative resource detailing spacecraft anomalies/failures & other related topics such as space weather events & on-orbit breakups; its limitations lie primarily in insufficient analysis/evaluation of root causes, a lack of comprehensive risk assessment, limited exploration of available tools/resources, potentially sensitive contact information disclosure, and absence of external perspectives. By addressing these concerns, the document could become an invaluable tool for improving our understanding of space operations and enhancing mission success rates.

MAGA PERSPECTIVE

This document is just another example of the liberal agenda pushing fear and paranoia about space anomalies and failures. It seems like they are trying to convince us that space missions are constantly at risk and that we need to rely on big government to protect us. Well, I for one am not buying it.

First of all, space weather events and on-orbit breakups have been happening since the beginning of space exploration. It's nothing new or unexpected. So why are they trying to make it seem like it's such a big deal? Maybe because they want us to be afraid and dependent on them for our safety.

And what about these so-called case studies? They are cherry-picking specific incidents to make it seem like failures and accidents are more common than they actually are. The truth is, space missions have been incredibly successful overall, with only a few isolated incidents.

Furthermore, the emphasis on understanding and mitigating the hazards of the space environment is just another excuse for government regulation and control. They want to limit our exploration and innovation in the name of safety. But true patriots know that taking risks is part of progress.

Finally, providing contact information for NASA personnel involved in this workshop only reinforces the idea that this document is meant to scare us into relying on them. They want us to call them for answers and guidance instead of thinking for ourselves.

In conclusion, this document is just propaganda meant to instill fear and dependence on big government. True Americans should see through their tactics and continue to boldly explore space without unnecessary regulations and interference.

PAGE-BY-PAGE SUMMARIES

and delayed crew rotation. Investigation ongoing to determine if there is a common cause for the anomalies.

BODY-13   A geomagnetic storm caused the loss of 38 Starlink satellites after a SpaceX launch, as the storm increased atmospheric drag. The satellites were put into safe-mode to minimize drag and updates were provided based on ground radars.

BODY-14   A minor geomagnetic storm caused the loss of 38 Starlink satellites deployed by SpaceX, resulting in increased atmospheric drag and the need for the satellites to enter safe mode.

BODY-15   Starlink Group 4-7 discusses the start of SpaceX Starlink launch operations during a period of low solar activity and the increasing solar and geomagnetic activity as we enter Solar Cycle 25.

BODY-16   Starlink Group 4-7 discusses the start of SpaceX's Starlink launch operations during a period of low solar activity and mentions that solar and geomagnetic activity are increasing as we enter Solar Cycle 25.

BODY-17   The page discusses the F107*Ap product as a measure of solar and geomagnetic activity's impact on atmospheric drag during Starlink launches. It mentions that the "experience envelope" for these conditions increased until 38 satellites were lost in Feb 2022.

BODY-18   The page discusses the effects of solar and geomagnetic activity on Starlink launches, including satellite losses and changes in orbit insertion altitude to reduce drag. Recent launches have returned to low altitude perigee but with higher apogee altitudes.

BODY-19   Starlink Group 4-7 experienced satellite losses due to high solar and geomagnetic activity, prompting SpaceX to change orbit insertion altitude. Recent launches have returned to low altitude perigee but with higher apogee altitudes. Group 6-1 launched during a strong geomagnetic storm.

BODY-20   Chandra X-Ray Observatory experienced performance degradation in its CCDs due to low energy radiation, but modified operating procedures successfully limited damage.

BODY-21   Protecting the operational life of ACIS with radiation interventions has resulted in 95 interruptions and 3197.9 hours of lost science time for the Chandra X-Ray Observatory.

BODY-23   The page lists the 100 largest X-ray flares from solar cycles 23, 24, and 25. These flares have low impact but can interfere with radio systems and satellite operations. They can also serve as a warning for geomagnetic storms and solar particle events.

BODY-24   This page provides information about Coronal Mass Ejections (CMEs), including their occurrence, velocity, and effects on geomagnetic storms and solar energetic particle events. It also mentions the collaboration between NASA, The Catholic University of America, and the Naval Research Laboratory in maintaining the CME catalog.

BODY-25   CMEs are increasing in number and speed, with the fastest ones having no impact on Earth. Data from NASA's CCMC Space Weather Database and StereoCaT tool provide information on CME properties.

BODY-27   The page provides information on the Kp index, which measures disturbances in the Earth's magnetic field. It explains that the Kp index exhibits a 27-day periodicity during solar minimum due to recurrent geomagnetic storms caused by high-speed solar wind flows.

BODY-28   The page provides information about the Kp index, which measures disturbances in the Earth's magnetic field. It explains that the Kp index exhibits a 27-day periodicity during solar minimum due to recurrent geomagnetic storms caused by high speed solar wind flows.

BODY-29   The page provides information on the Kp index, which measures geomagnetic activity. It explains that the index exhibits a 27-day periodicity during solar minimum but is affected by CME-driven storms during periods of increased solar activity. The page also mentions the appearance of Cycle 25 Kp in 2021.

BODY-30   The page provides information on the Kp index, a measure of geomagnetic activity. It explains how the index is affected by solar wind flows and coronal hole geometry, and mentions the appearance of Cycle 25 in 2021.

BODY-31   The page discusses the Kp index, which measures disturbances in the Earth's magnetic field. It explains that the index exhibits a 27-day periodicity during solar minimum, but this is washed out by CME driven storms during periods of increased solar activity. The page also mentions that Cycle 25 Kp ~ 7 periods first appeared in 2021.

BODY-32   Solar Cycles 20-23 had more frequent periods of geomagnetic activity with Kp ≥ 6, occurring at a rate of ~30 to 100 periods/year. However, Cycles 24 and 25 have been less active, with Kp ≥ 6 conditions occurring at a rate of only ~6 to 30 periods/year.

BODY-33   Solar Cycle 25 has been less active than previous cycles, with fewer periods of geomagnetic storms and no periods with Kp ≥ 8.

BODY-34   Satellite and debris populations are growing, posing a significant risk of collisions. The increase is attributed to events like Chinese ASAT test, accidental collision, and Russian ASAT test. The rise

in spacecraft numbers is due to satellite constellations, cubesats, and smallsats.

BODY-35   On-orbit breakups are a significant source of orbital debris, with over 250 objects fragmenting in low Earth orbit (LEO) in the last 60 years. Causes include explosions, collisions, and anti-satellite tests. Recent breakups include the Long March 6A upper stage in November 2022, generating hundreds of thousands of fragments. High altitude debris populations pose long-term threats to satellites.

BODY-36   The page provides statistics on the increasing number of payloads launched into Low Earth Orbit (LEO), with SpaceX being a major contributor. It also highlights the current population of objects in space, including functioning satellites and tracked debris.

BODY-37   There are over 9,000 objects in space, with around 7,200 still functioning. The number of satellites in low Earth orbit has increased significantly, especially in the commercial sector. If planned constellations are successful, the traffic and characteristics of the LEO operational environment could change drastically.

BODY-38   There are over 9,790 objects still in space, with around 7,200 of them still functioning. There are also estimated to be over 32,000 tracked and cataloged debris objects in orbit. The increase in payload launch traffic into LEO has been dominated by the commercial sector.

BODY-39   ISS has conducted 33 collision avoidance maneuvers since 1999, with the frequency depending on factors such as solar activity and the number of objects crossing its orbit. Despite numerous conjunctions with debris from the Cosmos 1408 ASAT test, only two maneuvers have been executed to avoid collisions.

NOTABLE PASSAGES

BODY-8   "Issues with new Starlink V2 design '...new technology in Starlink V2... experiencing some issues, as expected. Some sats will be deorbited, others will be tested thoroughly before raising altitude above Space Station.' E. Musk"

BODY-9   "Rocket failure (uncrewed science mission); fire in BE-3 engine, escape system carried capsule away from exploding rocket. Single BE-3PM booster engine suffered structural fatigue failure of nozzle, caused thrust misalignment that triggered the capsule's emergency escape system; 23rd launch."

BODY-10   Launch anomalies/failures typically result in loss of mission (reporting bias?) 14/15 missions, 48 payloads lost

BODY-11   "Anomalies and failure investigations and mitigation processes can be grouped into four broad categories: Initial stage: anomaly identified, cause(s) under investigation. Mid-stage: potential cause(s) identified, mitigation strategy(s) under development. Mature stage: cause(s) well understood, mitigation procedures in place. Failure: end of mission, cause(s) may have been identified, but no mitigation is possible."

BODY-13   "Unfortunately, the satellites deployed on Thursday [Feb 3] were significantly impacted by a geomagnetic storm on Friday. These storms cause the atmosphere to warm and atmospheric density at our low deployment altitudes to increase. In fact, onboard GPS suggests the escalation speed and severity of the storm caused atmospheric drag to increase up to 50 percent higher than during previous launches. The Starlink team commanded the satellites into a safe-mode where they would fly edge-on (like a sheet of paper) to minimize drag—to effectively 'take cover from the storm'—and continued to work closely with the Space Force's 18th Space Control Squadron and LeoLabs to provide updates on the satellites based on ground radars."

BODY-14   "Unfortunately, the satellites deployed on Thursday [Feb 3] were significantly impacted by a geomagnetic storm on Friday. These storms cause the atmosphere to warm and atmospheric density at our low deployment altitudes to increase. In fact, onboard GPS suggests the escalation speed and severity of the storm caused atmospheric drag to increase up to 50 percent higher than during previous launches. The Starlink team commanded the satellites into a safe-mode where they would fly edge-on (like a sheet of paper) to minimize drag—to effectively 'take cover from the storm'—and continued to work closely with the Space Force's 18th Space Control Squadron and LeoLabs to provide updates on the satellites based on ground radars."

BODY-19   "Following the Group 4-7 satellite losses, SpaceX altered their orbit insertion altitude to reduce the threat of increased drag at low altitudes during high solar and geomagnetic activity."

BODY-21   "Chandra science observations have been interrupted 95 times due to radiation events for a total of 3197.9 hours (~0.36 year) in its ~25 year operational life."

BODY-24   Fast CMEs are particularly geoeffective, driving geomagnetic storms and solar energetic particle events.

BODY-38   "Payload launch traffic into LEO has increased dramatically in recent years with the commercial sector dominating the increase. Traffic characteristics of the LEO operational environment could dramatically change in the coming years if even a fraction of the existing and proposed *constellations reach their targets for planned spacecraft."*

BODY-40 *"Collision avoidance maneuvers are a fact of life for low Earth orbit, is this a sign of the future for lunar operations as well as lunar exploration activities increase?"*

Artemis I (Image: NASA)

Spacecraft Anomalies and Failures Workshop 2023: NASA Introductory Comments

Joseph I. Minow, PhD
Technical Fellow for Space Environments
NASA Engineering and Safety Center
NASA, Marshall Space Flight Center

Spacecraft Anomalies and Failures 2023 Workshop
29 March 2023, GSFC, Greenbelt, MD
joseph.minow@nasa.gov

1

Outline

- Introduction and SCAF background

- Spacecraft, launch vehicle anomalies and failures in 2022 and 2023

- State of Solar Cycle 25 space environment

- Satellite traffic and orbital debris

- Summary

2

Introduction

- Welcome to NASA GSFC and SCAF 2023!

- Logistics
  - Fire, weather, restrooms
  - Maps to cafeteria at check-in desk

- Session chairs
  - Day 1: Joe Minow   NASA/MSFC
  - Day 2: Dong Ryu   NRO

- Organizing committee
  - Mike Campola   NASA/GSFC
  - Martha Obryan   GSFC/SSAI
  - Linda Parker   MSFC/Space Weather Solutions
  - Mike Squire   NASA/LARC
  - Yihua Zheng   NASA/GSFC

- Let us know if you have questions or need help today!

SPACECRAFT ANOMALIES & FAILURES WORKSHOP
"Creating a Community Solution for Anomaly Attribution"

Co-Sponsored by NRO and NASA
MARCH 29-30, 2023

Locations:
DAY 1 (UNCLASSIFIED): NASA Goddard Space Flight Center
DAY 2 (CLASSIFIED): NRO HQ Westfields

Presentations run from 9:00 AM to 4:00 PM EDT.
Check-In begins at 8:00 AM

AGENDA TOPICS:
· Spacecraft Anomalies, Failures, and Operations.
· Space Environmental Effects and Debris.
· Anomaly Recovery Operations and Anomaly Investigations.

OBJECTIVES:
· Review & share lessons learned from spacecraft anomalies & failures.
· Improve tradecraft for anomaly attribution & root cause determination.
· Reinforce relationships in the Space Community that do not regularly interact.

NRO POC: Dong Ryu, ryudong@nro.mil, 703-460-3697
NASA POC: Joseph Minow, joseph.minow@nasa.gov, 256-544-2850

SCAF Workshop History

- SCAF 2023 is the 10th in the SCAF workshop series that started in 2013

- Dr. Darren McKnight, Technical Director, Integrity Applications Incorporated (IAI) developed the original SCAF Workshop concept as an IAI organized event in 2013

- Workshops organized as annual event starting in 2013 through 2022 (no workshop in 2020) using a two-day format with invited presentations:
  - Day 1 unclassified sessions
  - Day 2 classified sessions

- NRO sponsorship of SCAF started with the 2016 workshop and NASA was included as a co-sponsor beginning in 2018 with Dr. McKnight/IAI and NRO continuing to serve as the primary organizer for the workshop

- In addition to the domestic SCAF workshops, Dr. McKnight organized an international Spacecraft Environmental Anomalies and Failures (SEAF) Workshop held in Toulouse, France in 2017 and Los Angeles, CA in 2019

- NASA and NRO took over as the co-organizers in 2021 with NASA organizing and hosting Day 1 (unclassified) presentations with NRO continuing to organize the Day 2 (classified) workshop

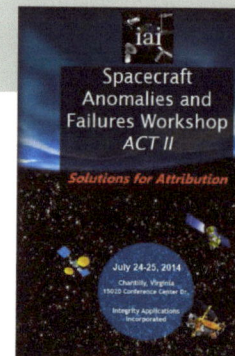

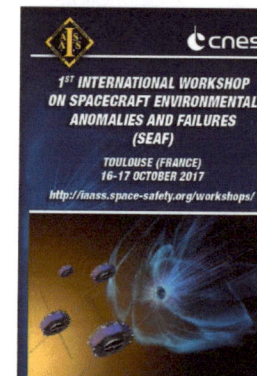

4

Spacecraft, Launch Vehicle Anomalies and Failures 2022-2023

- We are now 60+ years into the space age since the launch of the first spacecraft in 1957, space operations in many respects have become routine...

- However, anomalies and failures do continue to occur
  - Testing of new hardware
  - Aging of old hardware
  - Unexpected interactions with the space environment
  - In-space accidents and collisions

- Some failures occur early in the mission, others after many years on orbit

- The next few slides give examples of the range of anomalies and failures in both spacecraft and launch vehicles that occurred in 2022 and the first months of 2023
  - Wide range of environments
  - No one is immune: government, industry, academia
  - Public sources only

Sources:
- https://www.space.com/astra-rocket-loss-nasa-satellite-runaway-event
- https://www.space.com/nasa-geotail-magnetic-field-spacecraft-recorder-failed
- https://spacenews.com/intelsat-working-to-regain-control-of-galaxy-15-satellite/
- https://spacenews.com/progress-cargo-spacecraft-at-iss-suffers-coolant-leak/
- https://gizmodo.com/nasa-ibex-probe-operational-after-systems-reset-1850198754
- https://www.satellitetoday.com/in-space-services/2022/08/02/momentus-identifies-cause-of-vigoride-3-anomaly-deploys-4-more-satellites/
- https://interestingengineering.com/science/nasa-loses-contact-with-icon-satellite
- https://spacenews.com/virgin-orbit-narrows-down-cause-of-launcherone-failure/

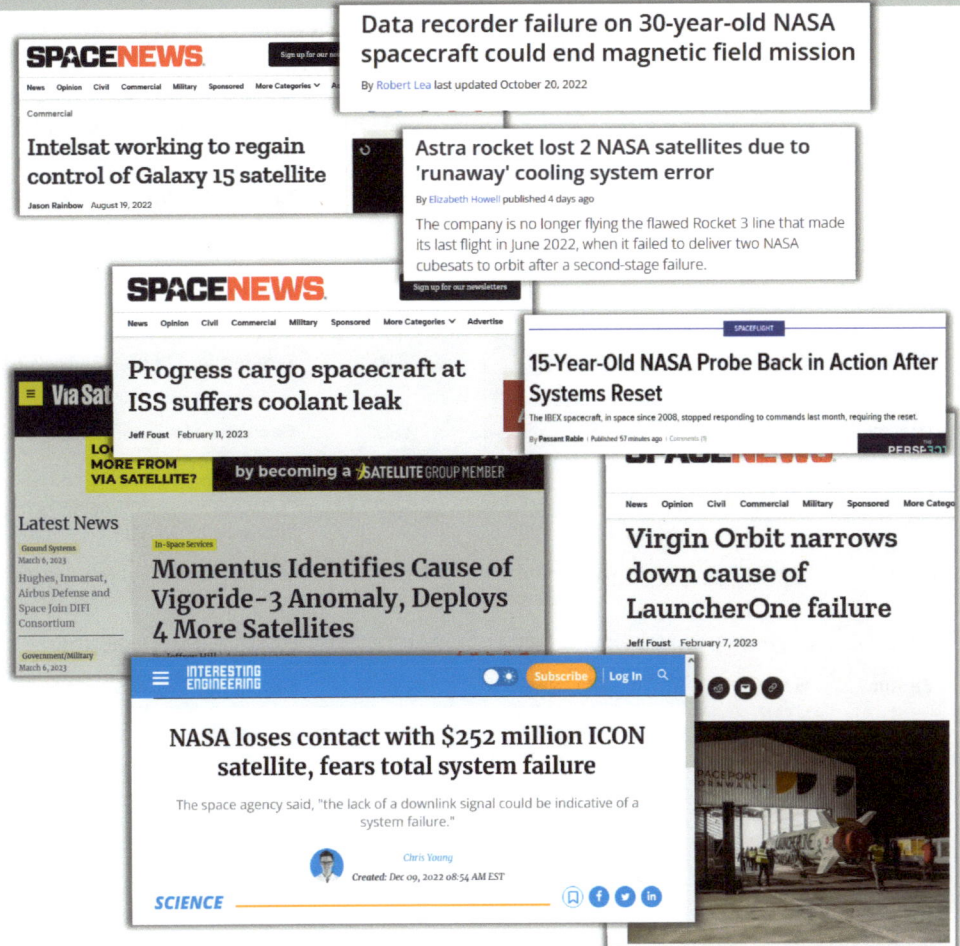

Data recorder failure on 30-year-old NASA spacecraft could end magnetic field mission
By Robert Lea last updated October 20, 2022

Intelsat working to regain control of Galaxy 15 satellite
Jason Rainbow August 19, 2022

Astra rocket lost 2 NASA satellites due to 'runaway' cooling system error
By Elizabeth Howell published 4 days ago
The company is no longer flying the flawed Rocket 3 line that made its last flight in June 2022, when it failed to deliver two NASA cubesats to orbit after a second-stage failure.

Progress cargo spacecraft at ISS suffers coolant leak
Jeff Foust February 11, 2023

15-Year-Old NASA Probe Back in Action After Systems Reset
The IBEX spacecraft, in space since 2008, stopped responding to commands last month, requiring the reset.
By Passant Rabie Published 57 minutes ago Comments (1)

Virgin Orbit narrows down cause of LauncherOne failure
Jeff Foust February 7, 2023

Momentus Identifies Cause of Vigoride-3 Anomaly, Deploys 4 More Satellites

NASA loses contact with $252 million ICON satellite, fears total system failure
The space agency said, "the lack of a downlink signal could be indicative of a system failure."
Chris Young
Created: Dec 09, 2022 08:54 AM EST

Selected Spacecraft Anomalies and Failures: 2022 and 2023

Date	Mission	Affiliation	Environment	Event	Cause/Notes	Ref*
5 Jan 22	ELSA-d	Astroscale	LEO	Anomalous spacecraft conditions required delay of orbital debris capture technology demo mission	Four of eight thrusters have technical difficulties and are now non-functional	1,2
3 Feb 22	Starlink	SpaceX	LEO	38 satellites lost	Loss of control, reentry due to high drag	3,4
4 Feb 22	SWIFT	NASA		Safe mode, February 2022	Failed reaction wheel, mechanical	5
31 Mar 22	Aqua	NASA	LEO	Safe mode 2022/090/17:59:08, Return to ops 2022/105/00:38	Unexpected power controller swap	6
22 Feb 22	MAVEN	NASA	Mars	Safe mode, 22 Feb – 28 May 2022	Inertial Measurement Unit anomaly during routine power cycle	7
19 May 22	Starliner OFT-2	NASA/Boeing	LEO	Two thrusters failed during orbital insertion burn	Primary and secondary units failed Tertiary backup operated successfully	8,9
25 May 22	JWST	NASA/ESA/CSA	Sun-Earth L2	Primary mirror damaged by hypervelocity impact between 23-25 May	Meteoroid impact on primary mirror	48,49
25 May 22	Vigoride-3	Momentus	LEO	Orbital transfer vehicle failure, failure to deploy 7 of 9 payloads	Solar arrays failed to deploy, reduced power and comm	10,11,52
20 Jun 22	Cygnus	NASA/NG	LEO	ISS reboost test ends early	Cygnus engine shut down 5 sec into 5 min burn	12
28 Jun 22	Geotail	JAXA/NASA	10x30 Re	Remaining data recorder failure, 28 June 2022	Shut down 28 Nov 22 after 30 years operations	14,47
5 Jul 22	CAPSTONE	NASA	Cislunar	Safe mode	Improperly formatted command	15
19 Aug 22	Galaxy 15	Intelsat	GEO	Loss of comm and control, drifting	Geomagnetic storm, broadcast payloads shut down as satellite drifts out of control	17,18, 19,55,56
24 Aug 22	JWST	NASA/ESA/CSA	Sun-Earth L2	Mid-Infrared Instrument grating wheel anomaly	Increased friction on grating wheel mechanism	20

*References in backup

6

Selected Spacecraft Anomalies and Failures: 2022 and 2023

Date	Mission	Affiliation	Environment	Event	Cause/Notes	Ref*
8 Sep 22	CAPSTONE	NASA	Cislunar	Tumble during trajectory correction maneuver	Vehicle attitude rates growing beyond the capacity of the onboard reaction wheels to control, propulsion system valve	15,16
10 Oct 22	TESS	NASA	LEO	Safe mode 10-13 Oct 22	Unexpected flight computer reset, recovered	22
25 Nov 22	ICON	NASA	LEO	Loss of contact with spacecraft 25 Nov 22, not recovered as of 10 Feb 23	Unknown; operating in extended mission since completing 2-yr prime mission science objectives	23,24
16 Nov 22	Orion (Artemis I)	NASA/ESA	Cislunar	RAM error in star tracker system, power conditioning and distribution unit (PCDU) fault	Thruster glow impact star tracker PCDU umbilical latching current limiter opened without command	25,26
16 Nov 22	Artemis I CubeSats	NASA	Cislunar	No comm with 5 of 10 deployed cubesats	LunaHMap (thruster failure), OMOTENASHI (no stable comm), NEAScout (no comm), CuSP (comm failed), LunIR (weak comm)	27
3 Nov 22	Orion (Artemis I)	NASA	Cislunar	Loss of comm with Orion for 47 minutes		28
7 Nov 22	Cygnus	NASA/NG	LEO	1 of 2 solar arrays fail to deploy	Vehicle successfully arrived at ISS to complete cargo mission	13
11 Dec 22	Orion (Artemis I)	NASA	LEO	Unpredicted loss of thermal protection (reentry heat shield) material	Vehicle survived reentry, but heat shield lost more charred material than expected	29
14 Dec 22	Lunar Flashlight	NASA	Cislunar	Reduced thruster performance	Fuel line obstruction	30
14 Dec 22	Soyuz MS-22	RSA	LEO	Coolant leak from thermal control system while docked to ISS	0.8 mm hole, meteoroid or orbital debris impact, Geminid meteor eliminated as cause	31,32,33, 34,35
3 Jan 23	Falcon-9/ Transporter-6	SpaceX/ USSF (US)	LEO	EWS RROCI spacecraft failed to deploy from launch vehicle, 113 of 114 payloads successfully deploy	Unclear if failure to deploy is spacecraft or launch vehicle failure; 195th flight	44,45,46

*References in backup

Date	Mission	Affiliation	Environment	Event	Cause/Notes	Ref*
3 Jan 23	Orbiter SN1	Launcher	LEO	Loss of power and control; failure to deploy 6 payloads, 4 hosted payloads lost	Orientation issue due to GPS antennae fault resulted in loss of power	37,38,53, 54
15 Jan 23	JWST	NASA/CSA	L2	NIRISS instrument comm delay, software timeout	GCR upset in FPGA, recovered	21
Mid Jan 23	SWOT	NASA	LEO	Ka-band Radar Interferometer (KaRIn) instrument high-power amplifier unexpectedly shut down		39,40
11 Feb 23	Progress MS-21	RSA	LEO	Coolant leak from thermal control system while docked to ISS	12 mm hole caused by "external influences"	36
16 Feb 23	MAVEN	NASA	Mars	Safe mode 16-17 Feb 22	Inertial Measurement Unit anomaly	41
18 Feb 23	IBEX	NASA	86,000 km x 260,000 km x 10.99°	Unresponsive to commands following flight computer reset	External reset successful on 2 March 23, reestablishing control of spacecraft (2 days before IBEX scheduled for autonomous power reset)	42,43
27 Feb 23	Starlink 6-1	SpaceX	LEO	Issues with new Starlink V2 design	"…new technology in Starlink V2… experiencing some issues, as expected. Some sats will be deorbited, others will be tested thoroughly before raising altitude above Space Station." E. Musk	50,51

Notes:

- **End of mission:** Starlink (38), Geotail, Galaxy-15?, ICON? Artemis CubeSats (5), Soyuz MS-22, EWS RROCI, Orbiter SN1 (and 10 payloads)

- **Anomalies/failure attributed to space environment:** (6/30 missions)
    - Ionizing radiation   2 (Orion, JWST)
    - M/OD   2 (Soyuz, JWST)
    - Atmospheric drag   1 (Starlink: 38 satellites)
    - Charging   1 (Galaxy-15)

*References in backup

8

Selected Launch Vehicle* Anomalies and Failures: 2022 and 2023

Date	Launch Vehicle	Affiliation	Mission	Event	Cause/Notes	Payloads Lost	Ref**
10 Feb 22	Rocket 3.3 (LV0008)	Astra (US)	Orb	Upper stage engine ignition before fairing separation, vehicle tumbling after the off-nominal stage separation	Fairing separation mechanism fired in incorrect order resulted in electrical anomaly and fairing separation failure; Upper stage engine unable to use Thrust Vector System due to software issue; Rocket 3.3 third flight	4	1,2,3
13 May 22	Hyperbola 1	iSpace (China)	Orb	2nd stage failed to ignite? Attitude control system anomaly?	iSpace = Beijing Interstellar Glory Space Tech. Ltd; third flight (all failures)	1	1,4,5
12 Jun 22	Rocket 3.3 (LV0010)	Astra (US)	Orb	Upper stage engine failure	Engine ran out of fuel and shut down early; fifth flight Rocket 3.3, only 2 of 5 3.3 missions were successful, 3.3 cancelled	2	1,6
7 Aug 22	SSLV	ISRO (India)	Orb	Final Velocity Trimming Module (VTM) stage failure, payloads in unstable orbits and lost	Unexpectedly strong shock during separation of 2nd stage saturated guidance system accelerometers, triggered "salvage mode"; payload deployed into unacceptably low orbit; first flight	2	1,7,8
12 Sep 22	New Shepard	Blue Origin (US)	Suborb	Rocket failure (uncrewed science mission); fire in BE-3 engine, escape system carried capsule away from exploding rocket	Single BE-3PM booster engine suffered structural fatigue failure of nozzle, caused thrust misalignment that triggered the capsule's emergency escape system; 23rd launch	---	1,9,10,42,45
1 Oct 22	Alpha	Firefly (US)	Orb	Payloads inserted into incorrect orbits, reenter early	Provider declare launch success, but payloads decayed early; second flight	3	1,11,12
8 Oct 22	Skylark L	Skyrora (Scotland)	Suborb	Booster failure	First flight	---	1,13
11 Oct 22	Epsilon	JAXA (Japan)	Orb	3rd stage ignition failure	Attitude control anomaly during 2nd stage coast; first flight	8	1,14,15,16,41
4 Nov 22	Long March 6A	China	Orb	Upper stage breakup, uncontrolled reentry	Successfully deployed Yunahi 3 satellite, subsequent breakup event initially released 50+ debris objects, over 350 cataloged by December; second flight LM 6A (LM 6 launched 10 successfully)	0	1,17,18,39,40

*Does not include pre-launch delays ("scrubs") or suborbital missile tests   **References in backup

9

Selected Launch Vehicle* Anomalies and Failures: 2022 and 2023

Date	Launch Vehicle	Affiliation	Mission	Event	Cause/Notes	Payloads Lost	Ref**
14 Dec 22	Zhuque-2	Landspace (China)	Orb	Early shutdown of 2nd stage engine	2nd stage vernier engine anomaly; first flight	14	1,19,20,21
20 Dec 22	Vega-C	Arianespace (France)	Orb	2nd stage engine anomaly	Under pressure in 2nd stage engine following launch and trajectory deviation, nozzle erosion; second flight	2	1,22, 23,36, 38
9 Jan 23	LauncherOne	Virgin Orbit (US)	Orb	2nd stage engine failure (Boeing 747 launch)	Filter dislodged in engine ($100 part); 6th flight	9	24,25,26, 37
10 Jan 23	RS1	ABL Space Systems (US)	Orb	Premature shutdown of all 9 of the RS1 first-stage engines at T+11 sec	Fire in RS1 avionics system, total electrical failure triggered engine shutdown; first flight	2	27,28,29, 30
7 Mar 23	H3	JAXA (Japan)	Orb	H3 rocket fails following launch, 2nd engine fails to ignite	H3 first test flight	1	34,35
22 Mar 23	Terran-1	Relativity Space	Orb	2nd stage engine anomaly	First 3-D printed rocket; first flight	---	43,44

Notes:

- Launch anomalies/failures typically result in loss of mission (reporting bias?)   14/15 missions, 48 payloads lost

- Many LV anomalies/failures occur early in program development:
  - 1st flight   7
  - 2nd flight   3
  - 3rd flight   2
  - 5th flight   1
  - 6th flight   1
  - >10 flights   1 (New Shephard: 23rd flight)

  - Total   15

*Does not include pre-launch delays ("scrubs") and suborbital missile tests   **References in backup

Anomalies and Failures

- The term "anomaly" often has a negative connotation for a spacecraft operator, in many cases the "anomaly" turns out to be:
  - Unexpected or unpredicted space environment interactions
  - Accidental: *collision, design flaws leading to breakup*
  - Deliberate: ASAT test, EOL destruct
  - Engineering failures that could have been prevented: *well, these <u>are</u> truly design or ops failures*

- Anomalies and failure investigations and mitigation processes can be grouped into four broad categories:
  - **Initial stage:** anomaly identified, cause(s) under investigation
  - **Mid-stage:** potential cause(s) identified, mitigation strategy(s) under development
  - **Mature stage:** cause(s) well understood, mitigation procedures in place
  - **Failure:** end of mission, cause(s) may have been identified, but no mitigation is possible

- I have three examples to show that demonstrate these stages:
  - Soyuz radiator anomaly (initial)
  - Starlink satellite losses (mid)
  - Chandra X-ray Observatory radiation interrupts (mature)

Robotic arm surveys Soyuz MS-22 after a radiator leak (NASA TV)

https://www.nasaspaceflight.com/2022/02/starlink-geostorm/

Chandra X-ray Observatory

NASA

11

Soyuz, Progress Coolant Leaks

Soyuz MS-22

«Союз МС-22»   Roscosmos

Изображение отверстия
в радиаторе системы
терморегулирования корабля,
сделанное камерами
манипулятора американского
сегмента МКС.

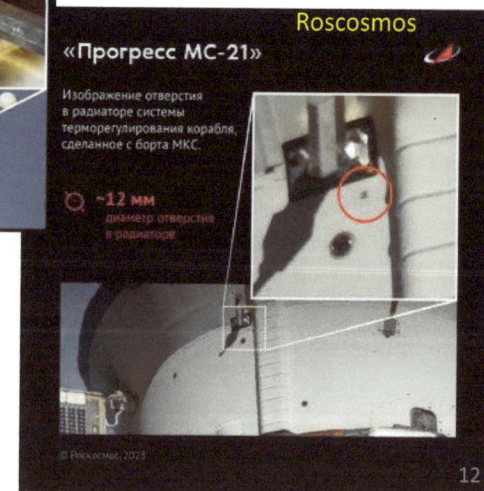

«Прогресс МС-21»   Roscosmos

Изображение отверстия
в радиаторе системы
терморегулирования корабля,
сделанное с борта МКС.

~12 мм
диаметр отверстия
в радиаторе

- Soyuz and Progress vehicles both experienced coolant leaks in radiator system on the spacecraft service module while docked to ISS
  - Soyuz MS-22 (crew)   14 Dec 2022   0.8 mm hole
  - Progress MS-21 (cargo)   11 Feb 2023   12 mm hole

- Roscosmos indicated the cause for the leaks was
  - Soyuz   meteoroid or orbital debris
  - Progress   "external influences" but not necessarily MMOD

- Mission impacts:
  - Russian EVA scheduled 14-15 Dec cancelled (crew suited up and waiting in airlock to egress when notified)
  - Crew rotation delayed while new Soyuz MS-23 launched to ISS to replace MS-22

- Two similar anomalies in similar locations on similar vehicles within a period of 2 months is unusual!
  - Investigation continues, trying to understand if there is a common cause or other causal factor in radiator panel leaks
  - Ground tests planned to simulate the damage

Sources::
- https://www.space.com/soyuz-spacecraft-leak-photos-russia-space-agency
- https://spacepolicyonline.com/news/progress-ms-21-leak-due-to-external-impact/
- https://spaceflightnow.com/2023/02/21/russia-blames-progress-coolant-leak-on-external-influences-as-replacement-soyuz-rolls-to-launch-pad/
- https://spacenews.com/replacement-soyuz-arrives-at-space-station/

Starlink Group 4-7

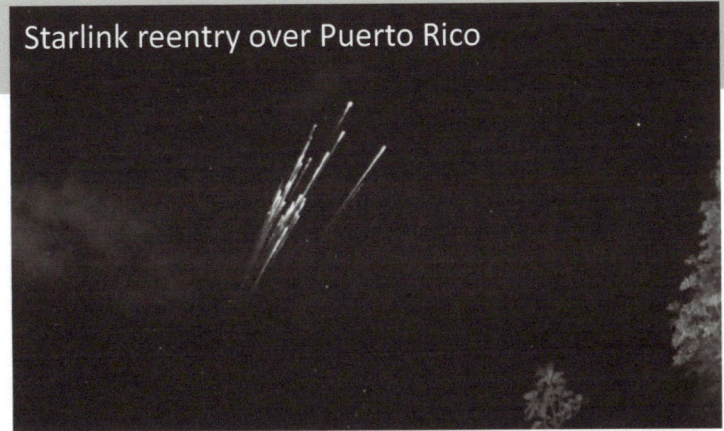

Starlink reentry over Puerto Rico

https://room.eu.com/news/spacex-starlink-satellites-fall-out-of-sky-after-storm

- SpaceX Group 4-7 launch of 49 Starlink satellites on 3 Feb 2022 encountered a minor geomagnetic storm on the day following launch, resulting in the loss of 38 Starlink satellites

SpaceX Update, February 8, 2022 (https://www.spacex.com/updates/)
Unfortunately, the satellites deployed on Thursday [Feb 3] were significantly impacted by a geomagnetic storm on Friday. These storms cause the atmosphere to warm and atmospheric density at our low deployment altitudes to increase. In fact, onboard GPS suggests the escalation speed and severity of the storm caused atmospheric drag to increase up to 50 percent higher than during previous launches. The Starlink team commanded the satellites into a safe-mode where they would fly edge-on (like a sheet of paper) to minimize drag—to effectively "take cover from the storm"—and continued to work closely with the Space Force's 18th Space Control Squadron and LeoLabs to provide updates on the satellites based on ground radars.

13

Starlink Group 4-7

- SpaceX Group 4-7 launch of 49 Starlink satellites on 3 Feb 2022 encountered a minor geomagnetic storm on the day following launch, resulting in the loss of 38 Starlink satellites

> ***SpaceX Update, February 8, 2022*** **(**https://www.spacex.com/updates/**)**
> ==*Unfortunately, the satellites deployed on Thursday [Feb 3] were significantly impacted by a geomagnetic storm on Friday. These storms cause the atmosphere to warm and atmospheric density at our low deployment altitudes to increase.*== *In fact, onboard GPS suggests the escalation speed and severity of the storm caused atmospheric drag to increase up to 50 percent higher than during previous launches.* ==*The Starlink team commanded the satellites into a safe-mode where they would fly edge-on (like a sheet of paper) to minimize drag*==*—to effectively "take cover from the storm"—and continued to work closely with the Space Force's 18th Space Control Squadron and LeoLabs to provide updates on the satellites based on ground radars.*

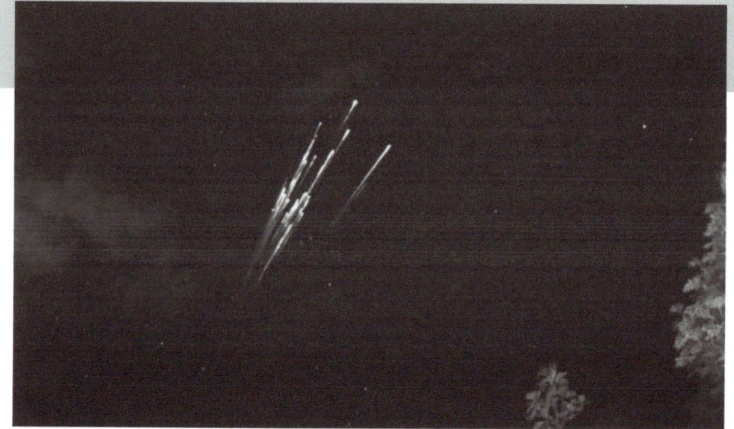

https://room.eu.com/news/spacex-starlink-satellites-fall-out-of-sky-after-storm

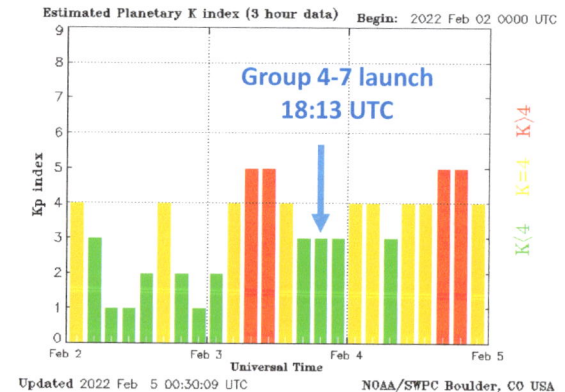

- Geomagnetic activity on 3 and 4 February were not that severe, at best only minor storm levels:
  - Maximum Kp never exceeded Kp 4 or 5 during the minor geomagnetic storms on both days
  - Hourly ap values never exceeded 56
  - Peak Dst was only -66 nT

Starlink Group 4-7

- SpaceX Starlink launch operations started during the solar minimum period between Solar Cycle 24 and 25

- Solar and geomagnetic activity are both increasing as we move into Solar Cycle 25

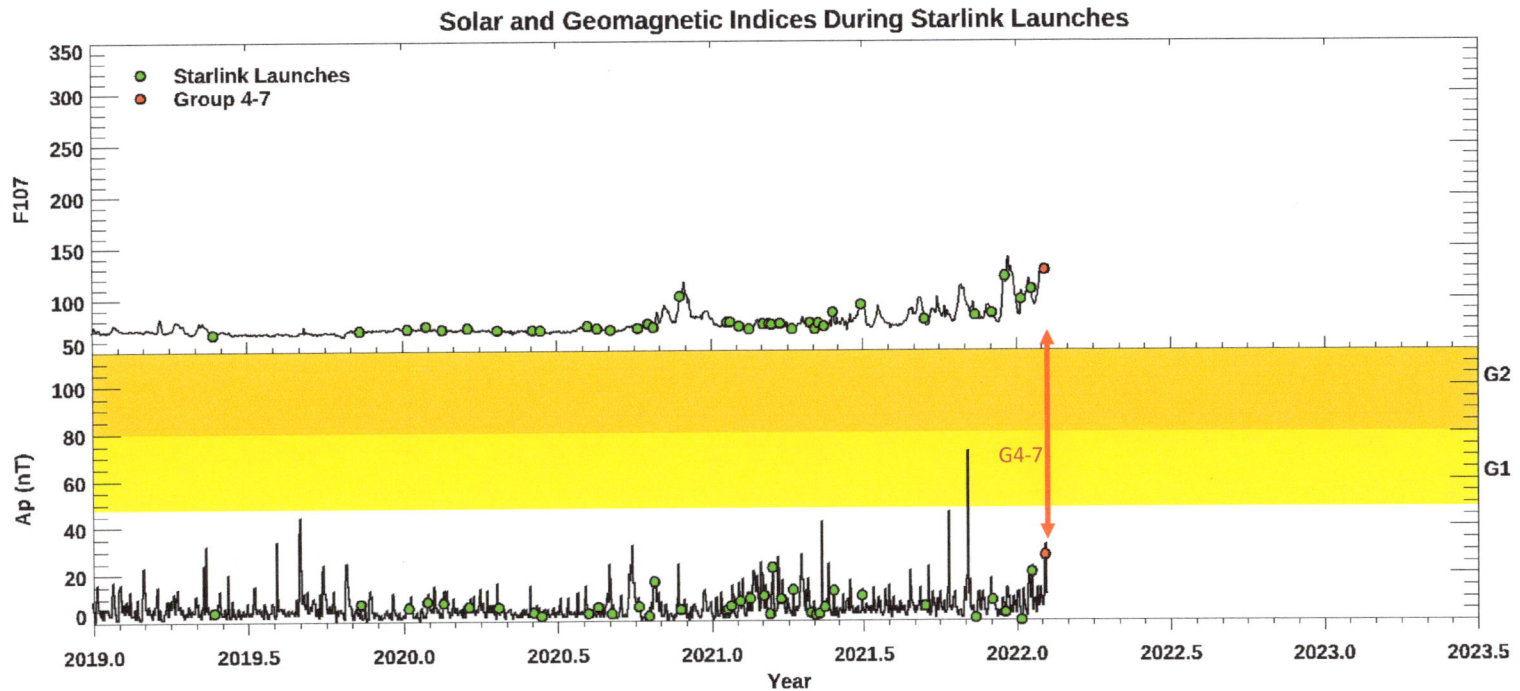

Solar and Geomagnetic Indices During Starlink Launches

Note: if 3-hour ap had been used instead of the daily average Ap, the hourly ap 48 and 56 values would have been in the low G1 range.

15

- SpaceX Starlink launch operations started during the solar minimum period between Solar Cycle 24 and 25

- Solar and geomagnetic activity are both increasing as we move into Solar Cycle 25

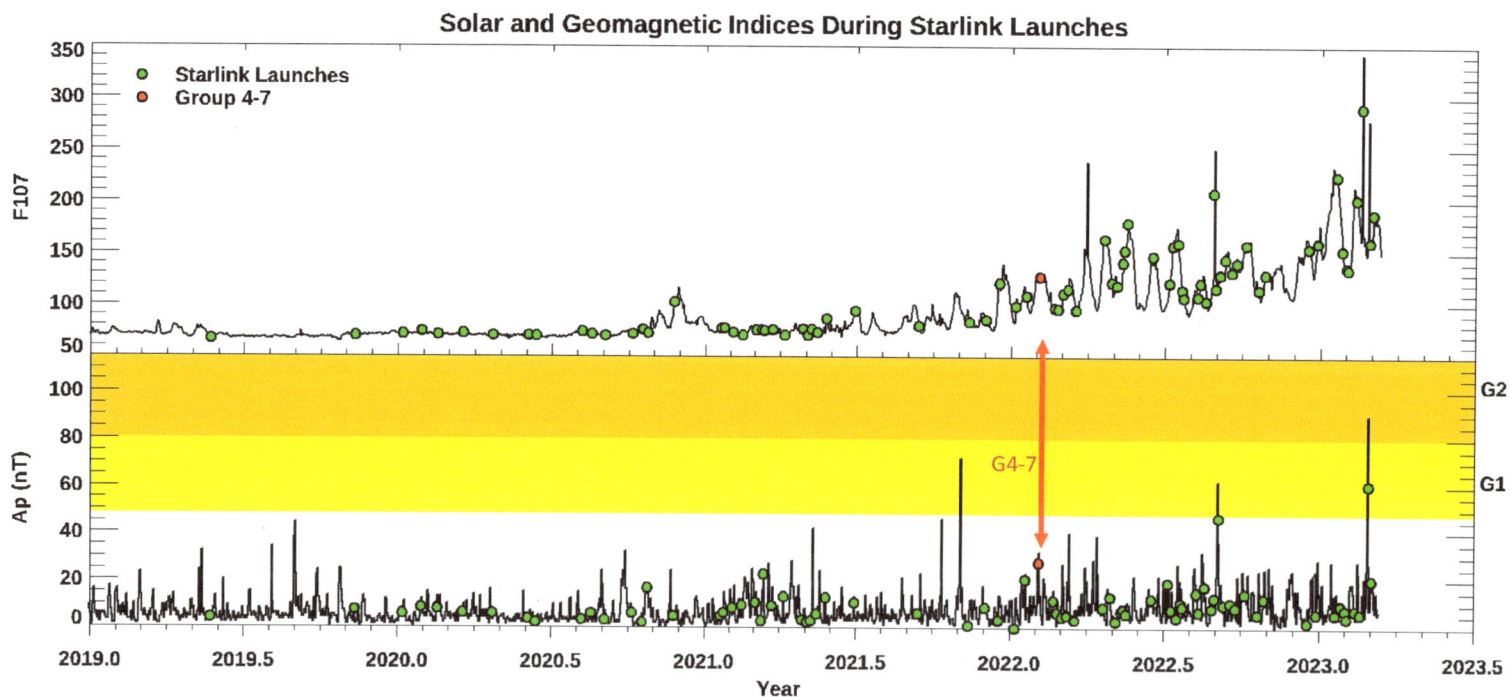

Solar and Geomagnetic Indices During Starlink Launches

Note: if 3-hour ap had been used instead of the daily average Ap, the hourly ap 48 and 56 values would have been in the low G1 range.

16

Starlink Group 4-7

- The F107*Ap product is a proxy for the combined effects of solar and geomagnetic activity, demonstrating an "experience envelope" for space weather conditions associated with atmospheric drag during Starlink launches

- Starlink "experience envelope" in solar and geomagnetic activity gradually increased until the Feb 2022 events where 38 satellites were lost

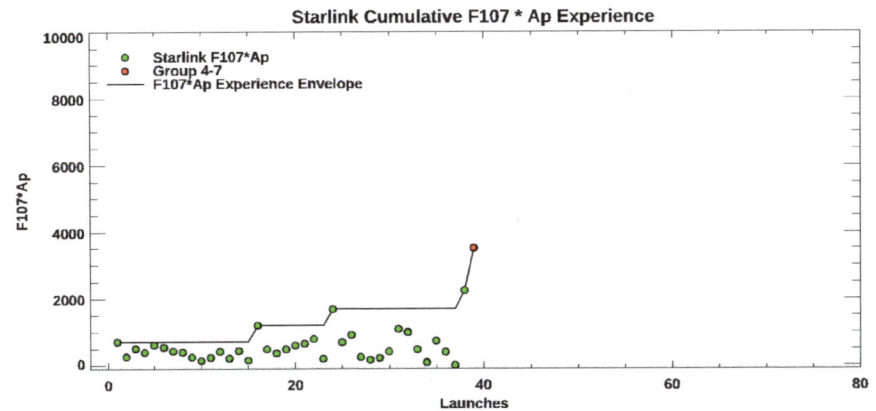

Starlink Cumulative F107 * Ap Experience

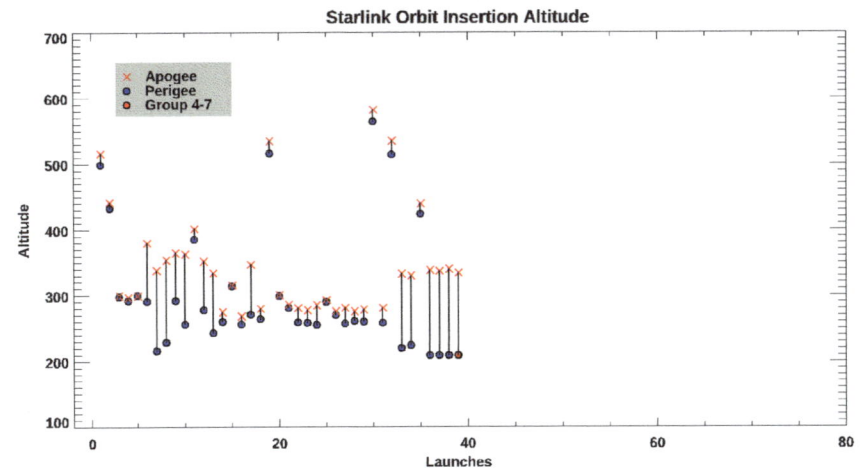

Starlink Orbit Insertion Altitude

https://www.nasaspaceflight.com/2023/03/starlink-6-1-2-7/
https://everydayastronaut.com/starlink-group-6-1-falcon-9-block-5-2/

Starlink Group 4-7

- The F107*Ap product is a proxy for the combined effects of solar and geomagnetic activity, demonstrating an "experience envelope" for space weather conditions associated with atmospheric drag during Starlink launches

- Starlink "experience envelope" in solar and geomagnetic activity gradually increased until the Feb 2022 events where 38 satellites were lost

- Following the Group 4-7 satellite losses, SpaceX altered their orbit insertion altitude to reduce the threat of increased drag at low altitudes during high solar and geomagnetic activity

- More recent launches are back to using low altitude perigee but the apogee altitude remain ≥ 300 km which does reduce the overall drag, however many insertion orbits are now similar to the one used for Group 4-7

Starlink Cumulative F107 * Ap Experience

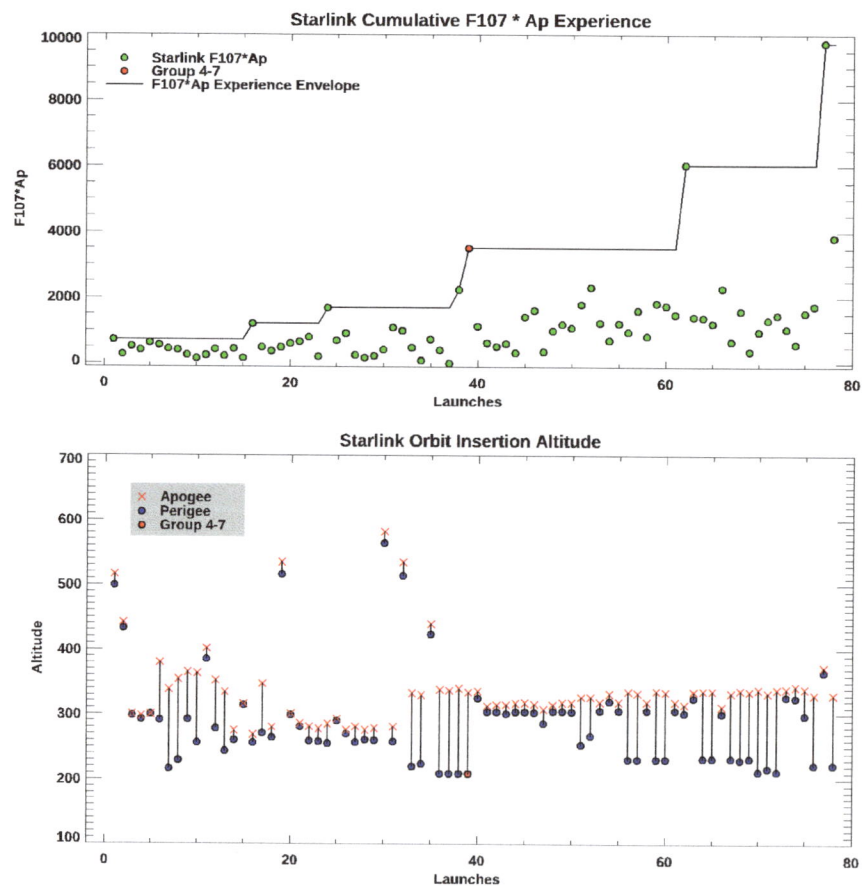

Starlink Orbit Insertion Altitude

https://www.nasaspaceflight.com/2023/03/starlink-6-1-2-7/
https://everydayastronaut.com/starlink-group-6-1-falcon-9-block-5-2/

18

Starlink Group 4-7

- The F107*Ap product is a proxy for the combined effects of solar and geomagnetic activity, demonstrating an "experience envelope" for space weather conditions associated with atmospheric drag during Starlink launches

- Starlink "experience envelope" in solar and geomagnetic activity gradually increased until the Feb 2022 events where 38 satellites were lost

- Following the Group 4-7 satellite losses, SpaceX altered their orbit insertion altitude to reduce the threat of increased drag at low altitudes during high solar and geomagnetic activity

- More recent launches are back to using low altitude perigee but the apogee altitude remain ≥ 300 km which does reduce the overall drag, however many insertion orbits are now similar to the one used for Group 4-7

- Group 6-1 launched 23:13 UT on 27 February 2023 at the end of a day with Kp = 6, 6+, and 7- geomagnetic storm conditions, <u>one of the strongest geomagnetic storms this solar cycle!</u> Launch deployed 21 Starlink v2-mini satellites at ~370 km with new spacecraft design.

https://www.nasaspaceflight.com/2023/03/starlink-6-1-2-7/
https://everydayastronaut.com/starlink-group-6-1-falcon-9-block-5-2/

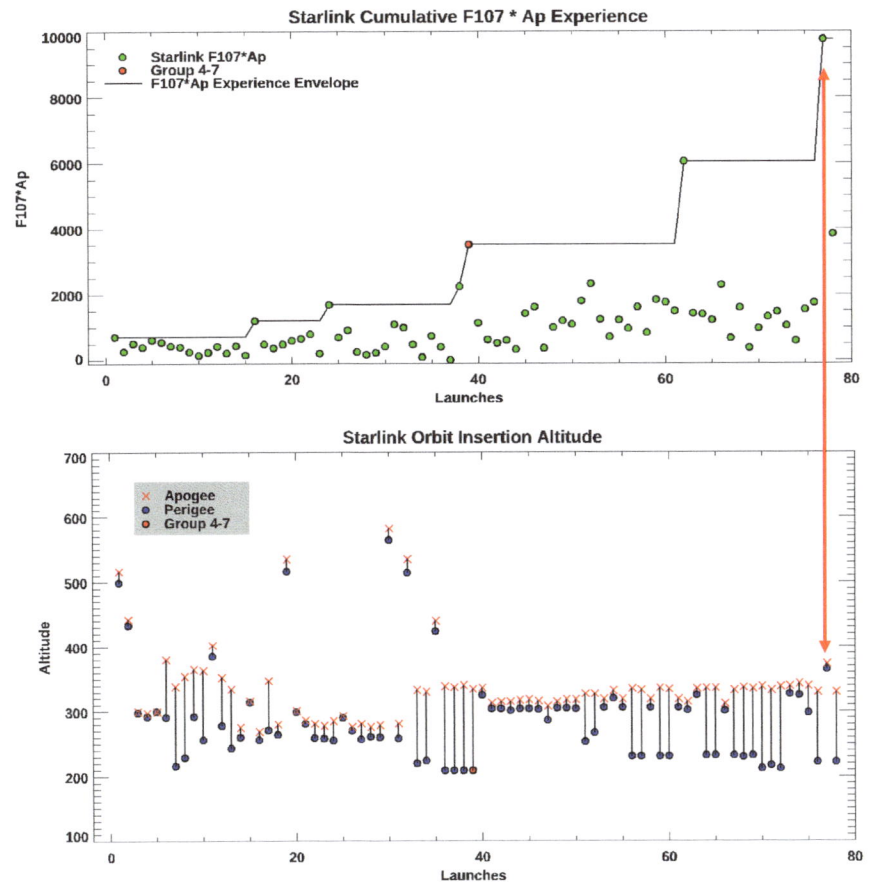

Starlink Cumulative F107 * Ap Experience

Starlink Orbit Insertion Altitude

19

Chandra X-Ray Observatory

- Launched July 1999, Chandra is now in the 24th year of operations
  Initial orbit   16,000 km x 133,000 km x 28.5° (~64 hr)
  Current orbit   1,538 km x 147,259 km x 38.8°

- The *Advanced CCD Imaging Spectrometer* (ACIS) instrument experienced an anomaly characterized by rapid performance degradation in front illuminated CCDs with no damage to back illuminated CCDs during initial flight operations in 1999

- Chandra identified weakly penetrating, low energy (0.1-0.5 MeV) protons present during radiation belt transits and solar particle events as the cause

- The low energy radiation damage mechanism was not predicted before flight

- Operating procedures were modified to move ACIS to a protected location out of the focal position during radiation belt passes, solar particle events

- While the radiation-protection program impacts science operations, it has been highly successful in limiting radiation damage to the ACIS CCDs to acceptable levels

2023/02/27   140.8 ks (39.1 hrs) manual

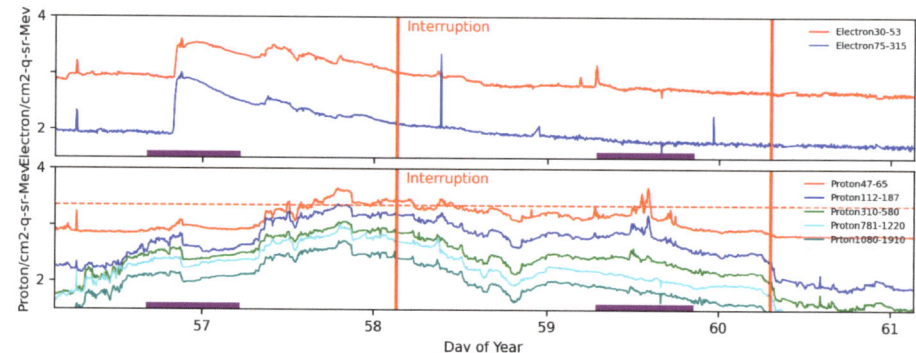

2021/10/28   197.7 ks (54.9 hrs) auto

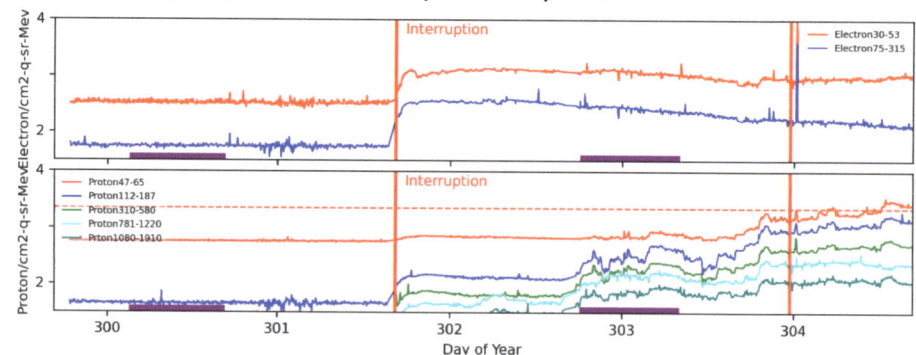

https://cxc.harvard.edu/mta/RADIATION/

20

Chandra X-Ray Observatory

- Protecting the operational life of ACIS with auto and manual radiation interventions comes at the cost of lost science time

- Chandra science observations have been interrupted 95 times due to radiation events for a total of 3197.9 hours (~0.36 year) in its ~25 year operational life

Solar Cycle	Number of Interrupts		Lost Science Time (hours)	
	auto	manual	auto	manual
23	45	20	1556.3	528.2
25	14	12	619.0	244.6
25	2	2	181.6	68.2
Total	61	34	2356.9	841.0

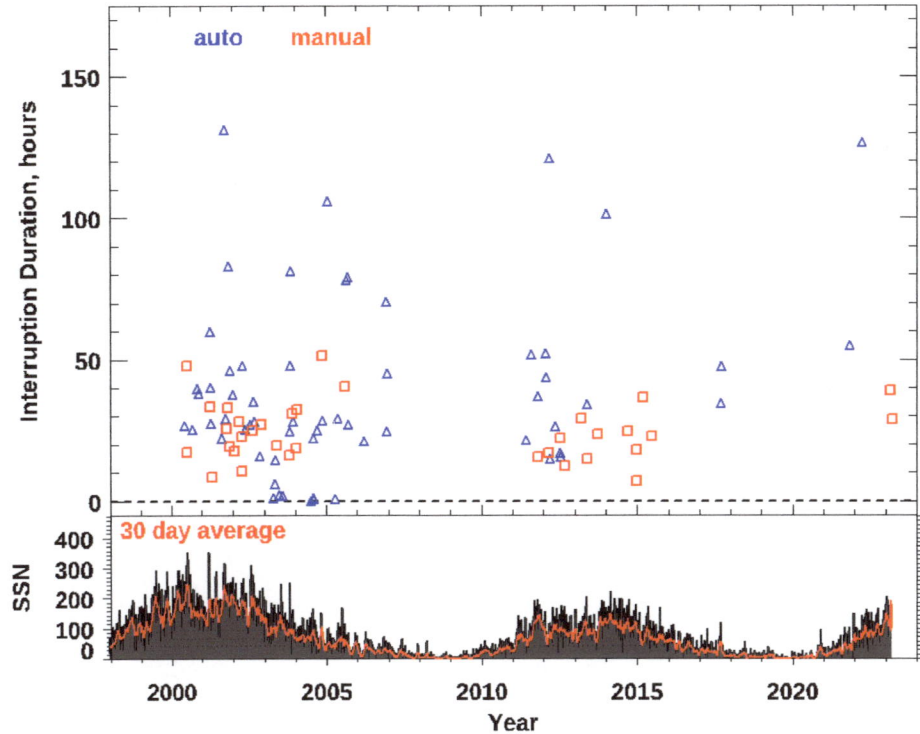

Chandra Radiation Interruptions

Sources:
- O'Dell et al., Managing radiation degradation of CCDs on the Chandra X-ray Observatory III, *Proc. SPIE 6686, UV, X-Ray, and Gamma-Ray Space Instrumentation for Astronomy XV*, 668603 (18 September 2007) https://doi.org/10.1117/12.734594
https://ntrs.nasa.gov/api/citations/20070039074/downloads/20070039074.pdf
- https://cxc.harvard.edu/mta/RADIATION/
- https://chandra.harvard.edu/about/tracking.html
- https://www.space-track.org/

21

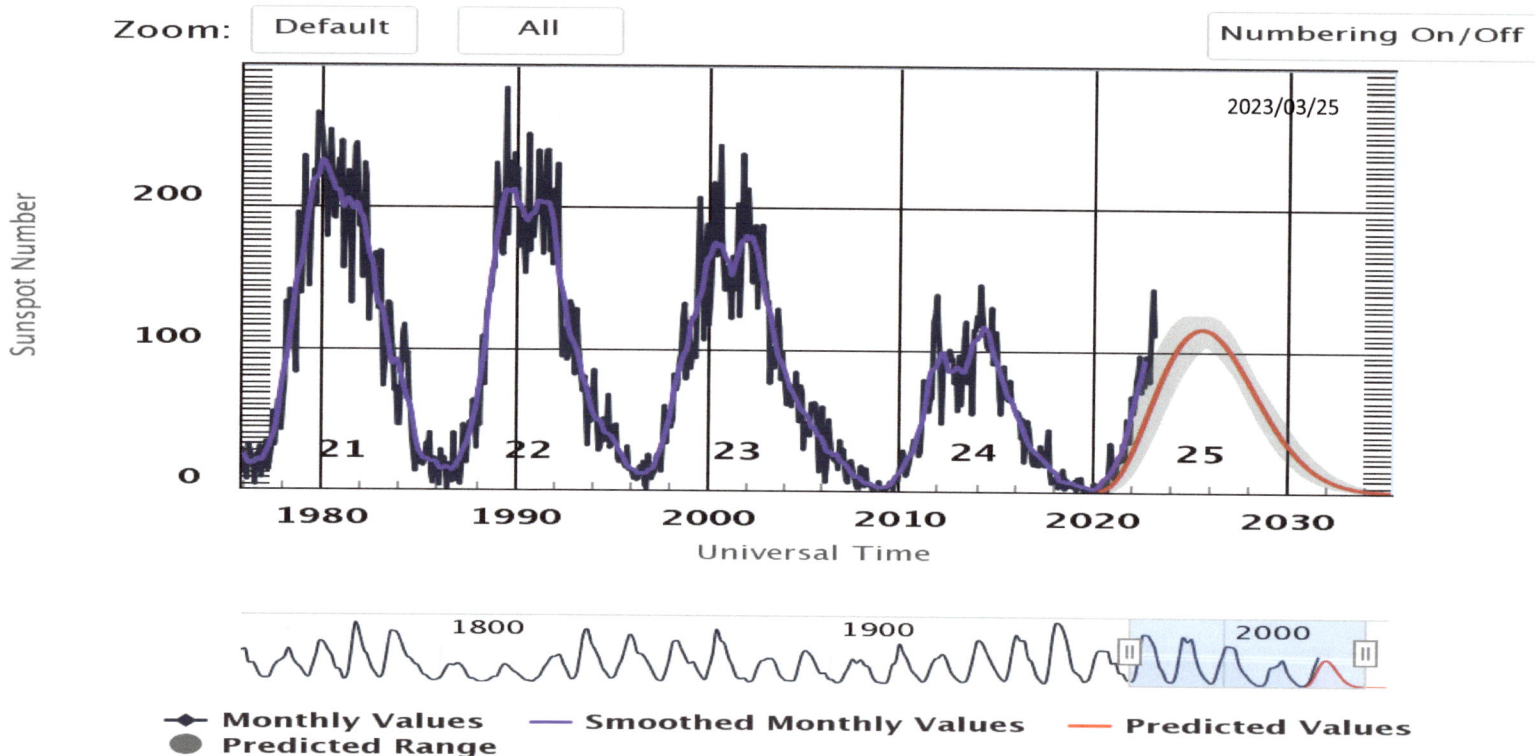

Solar Activity

ISES Solar Cycle Sunspot Number Progression

Zoom: Default | All   Numbering On/Off

2023/03/25

Monthly Values ◆ — **Smoothed Monthly Values** — **Predicted Values**
● **Predicted Range**

Space Weather Prediction Center

https://www.swpc.noaa.gov/products/solar-cycle-progression

Solar X-Ray Flares

- Major SWx impact of x-ray flares is increased ionization in the D-region ionosphere which interferes with terrestrial HF radio systems
- Large M and X class flares are correlated with coronal mass ejections and solar particle events (SPE) which can potentially impact satellite operations
- X-ray flares can provide advanced warning for geomagnetic storms, SPEs

X-Ray Flares, Solar Cycles 23, 24, 25

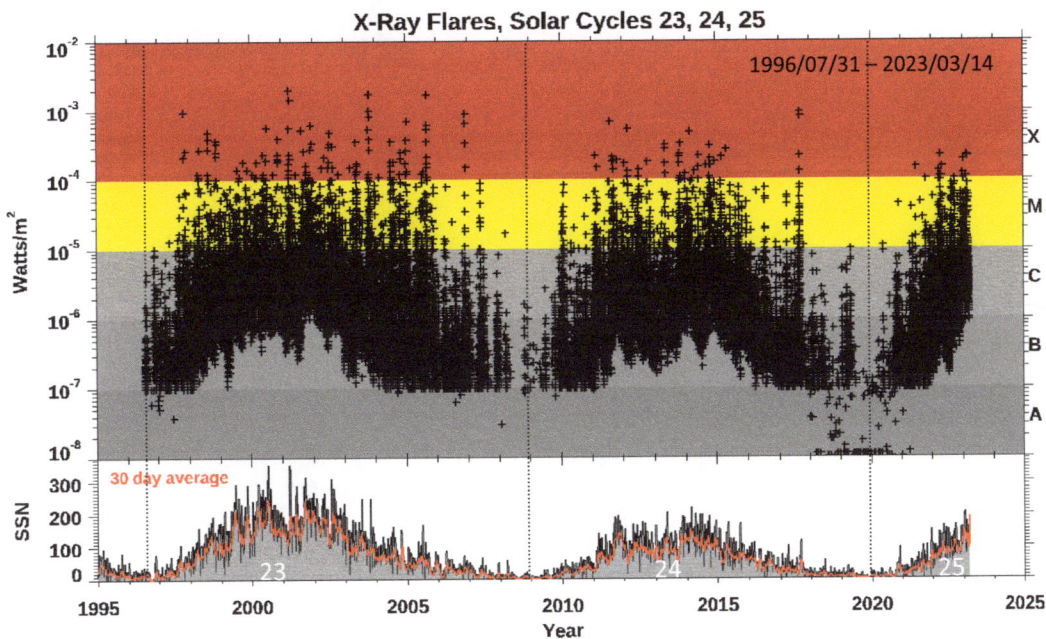

1996/07/31 – 2023/03/14

X-ray flare data source:
https://www.swpc.noaa.gov/products/solar-and-geophysical-event-reports

100 Largest X-Ray Flares in Cycles 23, 24, and 25

Rank	YYYY DOY HH:MM	Class	Cycle	Rank	YYYY DOY HH:MM	Class	Cycle	Rank	YYYY DOY HH:MM	Class	Cycle
1	2001 092 21:51	X20	23	39	2013 134 01:11	X3.2	24	77	2011 307 20:27	X1.9	24
2	2003 308 19:53	X17.4	23	40	2002 236 01:12	X3.1	23	78	2011 267 09:40	X1.9	24
3	2003 301 11:10	X17.2	23	41	2014 297 21:41	X3.1	24	79	2000 194 10:37	X1.9	23
4	2005 250 17:40	X17.0	23	42	2002 196 20:08	X3.0	23	80	2000 330 18:44	X1.9	23
5	2001 105 13:50	X14.4	23	43	2013 133 16:05	X2.8	23	81	2014 354 00:28	X1.8	24
6	2003 302 20:49	X10.0	23	44	1998 230 08:24	X2.8	23	82	2000 329 21:59	X1.8	23
7	1997 310 11:55	X9.4	23	45	2001 345 08:08	X2.8	23	83	2004 231 17:40	X1.8	23
8	2017 249 12:02	X9.3	24	46	2003 307 01:30	X2.7	23	84	2012 297 03:17	X1.8	24
9	2006 339 10:35	X9.0	23	47	1998 126 08:09	X2.7	23	85	1999 287 09:00	X1.8	23
10	2003 306 17:25	X8.3	23	48	2015 125 22:11	X2.7	24	86	2002 199 07:44	X1.8	23
11	2017 253 16:06	X8.2	24	49	1997 331 13:17	X2.6	23	87	2000 084 07:52	X1.8	23
12	2005 020 07:01	X7.1	23	50	2005 015 23:02	X2.6	23	88	2004 197 01:41	X1.8	23
13	2011 221 08:05	X6.9	24	51	2001 267 10:38	X2.6	23	89	2011 250 22:38	X1.8	24
14	2006 340 18:47	X6.5	23	52	1998 326 16:23	X2.5	23	90	2003 160 21:39	X1.7	23
15	2005 252 20:04	X6.2	23	53	2004 315 02:13	X2.5	23	91	2013 298 08:01	X1.7	24
16	2001 347 14:30	X6.2	23	54	2000 158 15:25	X2.3	23	92	2005 001 00:31	X1.7	23
17	2003 196 10:24	X5.7	23	55	2000 329 15:13	X2.3	23	93	2001 088 10:15	X1.7	23
18	2001 096 19:21	X5.6	23	56	2013 302 21:54	X2.3	24	94	2012 027 18:37	X1.7	24
19	2012 067 00:24	X5.4	24	57	2001 100 05:26	X2.3	23	95	2005 256 23:22	X1.7	23
20	2005 251 21:06	X5.4	23	58	2023 048 20:16	X2.2	25	96	2013 133 02:17	X1.7	24
21	2003 296 08:35	X5.4	23	59	2022 110 03:57	X2.2	25	97	2003 162 20:14	X1.6	23
22	2001 237 16:45	X5.3	23	60	2017 249 09:10	X2.2	24	98	2014 253 17:45	X1.6	24
23	1998 230 22:19	X4.9	23	61	2014 161 11:42	X2.2	24	99	2001 292 01:05	X1.6	23
24	2014 056 00:49	X4.9	24	62	1998 327 06:44	X2.2	23	100	2014 311 17:26	X1.6	24
25	2002 204 00:35	X4.8	23	63	2011 046 01:56	X2.2	24				
26	2000 331 16:48	X4.0	23	64	1997 308 05:58	X2.1	23				
27	2003 307 09:55	X3.9	23	65	2013 298 15:03	X2.1	24				
28	1998 231 21:45	X3.9	23	66	2023 062 17:52	X2.1	25				
29	2005 017 09:52	X3.8	23	67	2011 249 22:20	X2.1	24				
30	1998 326 06:42	X3.7	23	68	2015 070 16:22	X2.1	24				
31	2005 252 09:59	X3.6	23	69	2002 140 15:27	X2.1	23				
32	2003 148 00:27	X3.6	23	70	2005 253 22:11	X2.1	23				
33	2004 198 13:55	X3.6	23	71	2001 102 10:28	X2.0	23				
34	2006 347 02:40	X3.4	23	72	2004 312 16:06	X2.0	23				
35	2001 362 20:45	X3.4	23	73	2000 329 05:02	X2.0	23				
36	2002 201 21:30	X3.3	23	74	2014 300 14:47	X2.0	24				
37	1998 332 05:52	X3.3	23	75	2014 299 10:56	X2.0	24				
38	2013 309 22:12	X3.3	24	76	2023 009 18:50	X1.9	25				

Largest Cycle 25 x-ray flares to date have only been low X-class events

23

Coronal Mass Ejections (CME)

- CME linear speeds from the SOHO LASCO CME Catalog provides a good summary of CME occurrence and velocity as a function of phase in solar cycle
- Fast CMEs are particularly geoeffective, driving geomagnetic storms and solar energetic particle events
- Relativistic electron enhancements in the radiation belts (internal charging), magnetospheric hot plasma (surface charging), and solar protons and heavy ions (single event effects) are correlated with CME activity

https://soho.nascom.nasa.gov/gallery/Movies/flares.html

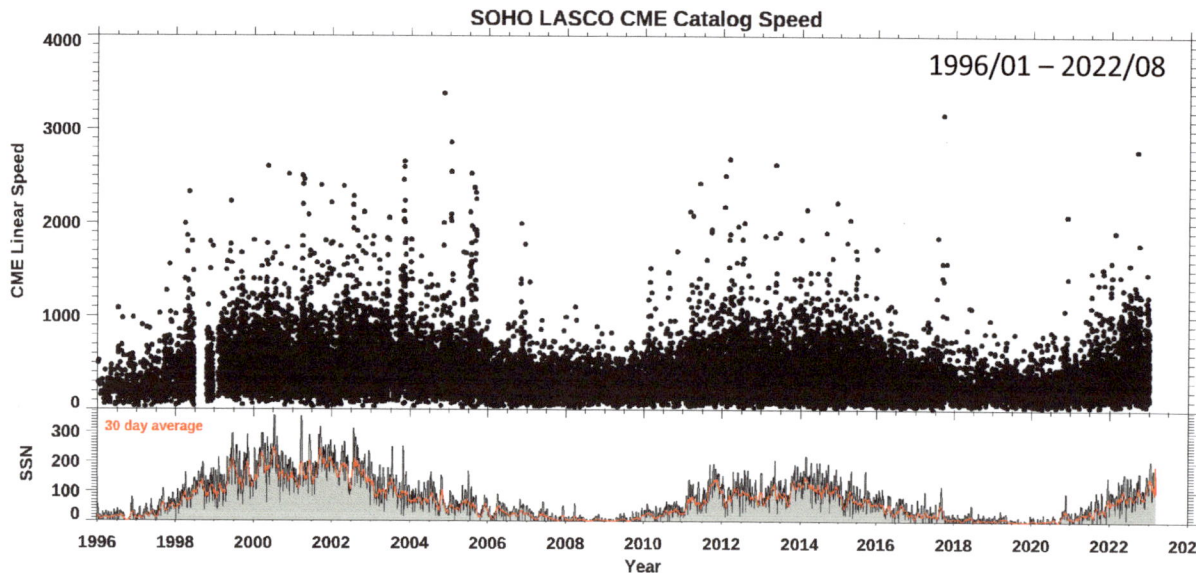

SOHO LASCO CME Catalog Speed

1996/01 – 2022/08

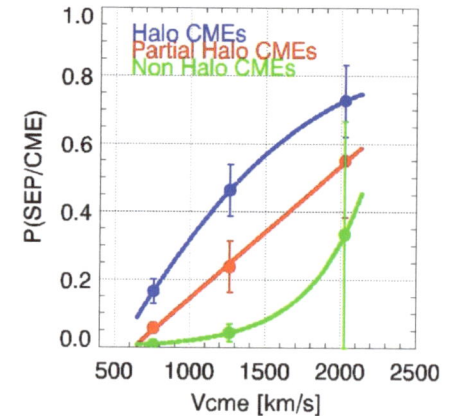

Papaioannou et al., 2018

Plane-of-sky velocity from SOHO LASCO images

URL: https://cdaw.gsfc.nasa.gov/CME_list/
- This CME catalog is generated and maintained at the CDAW Data Center by NASA and The Catholic University of America in cooperation with the Naval Research Laboratory.
- SOHO is a project of international cooperation between ESA and NASA.

SSN: GFZ German Research Centre for Geosciences

24

CME Initial Velocity: 2018 to present

- CMEs are increasing in number and larger number of high speed CMEs as solar activity increases
- Fastest CMEs (>2000 km/s) to date are farside events with no direct impact on near Earth space
- Data source: NASA/GSFC: CCMC *Space Weather Database of Notifications, Knowledge, Information* (DONKI)
  URL: https://ccmc.gsfc.nasa.gov/tools/DONKI/
  CME from Moon to Mars Catalog (most accurate only)

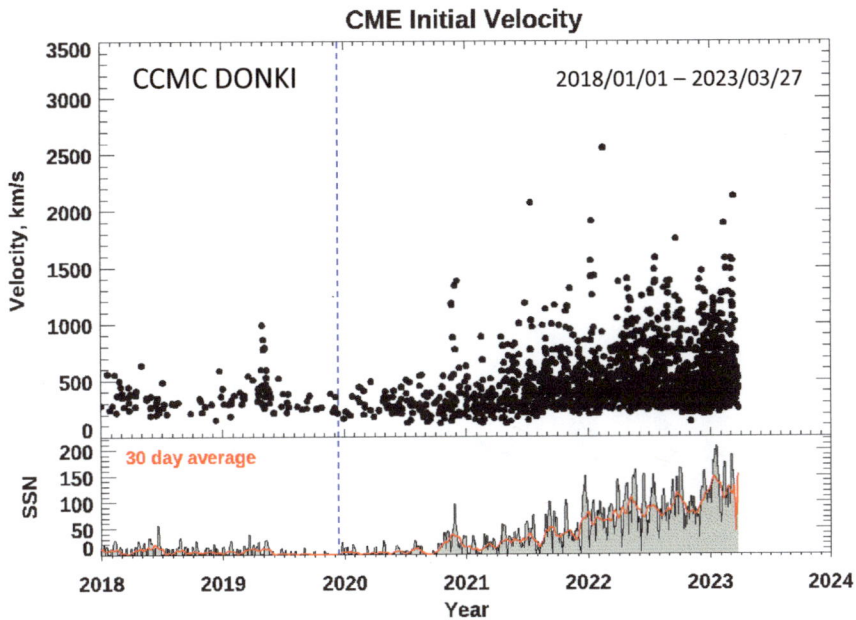

M2M estimates of CME 3-D kinematic properties (including velocity) are from StereoCaT tool

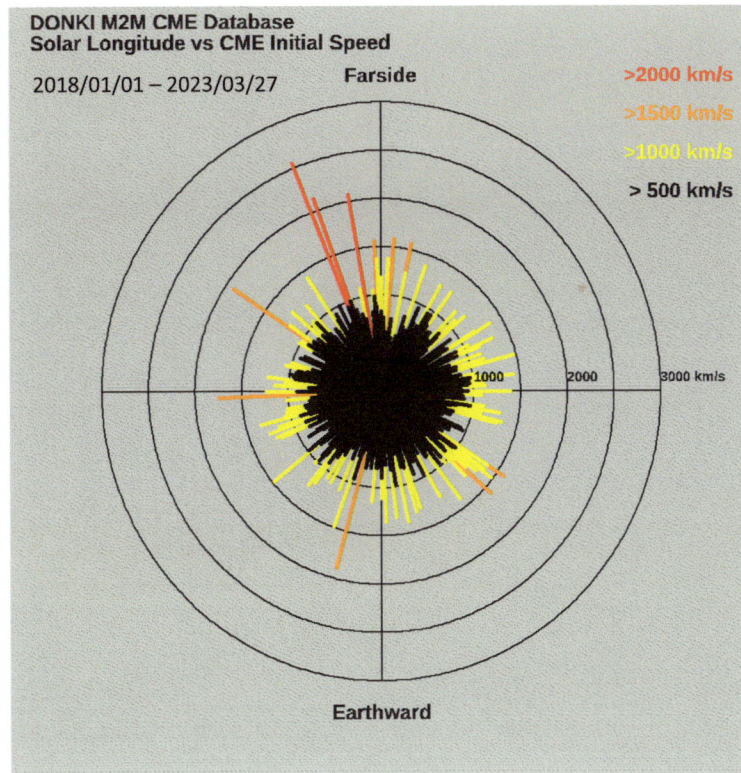

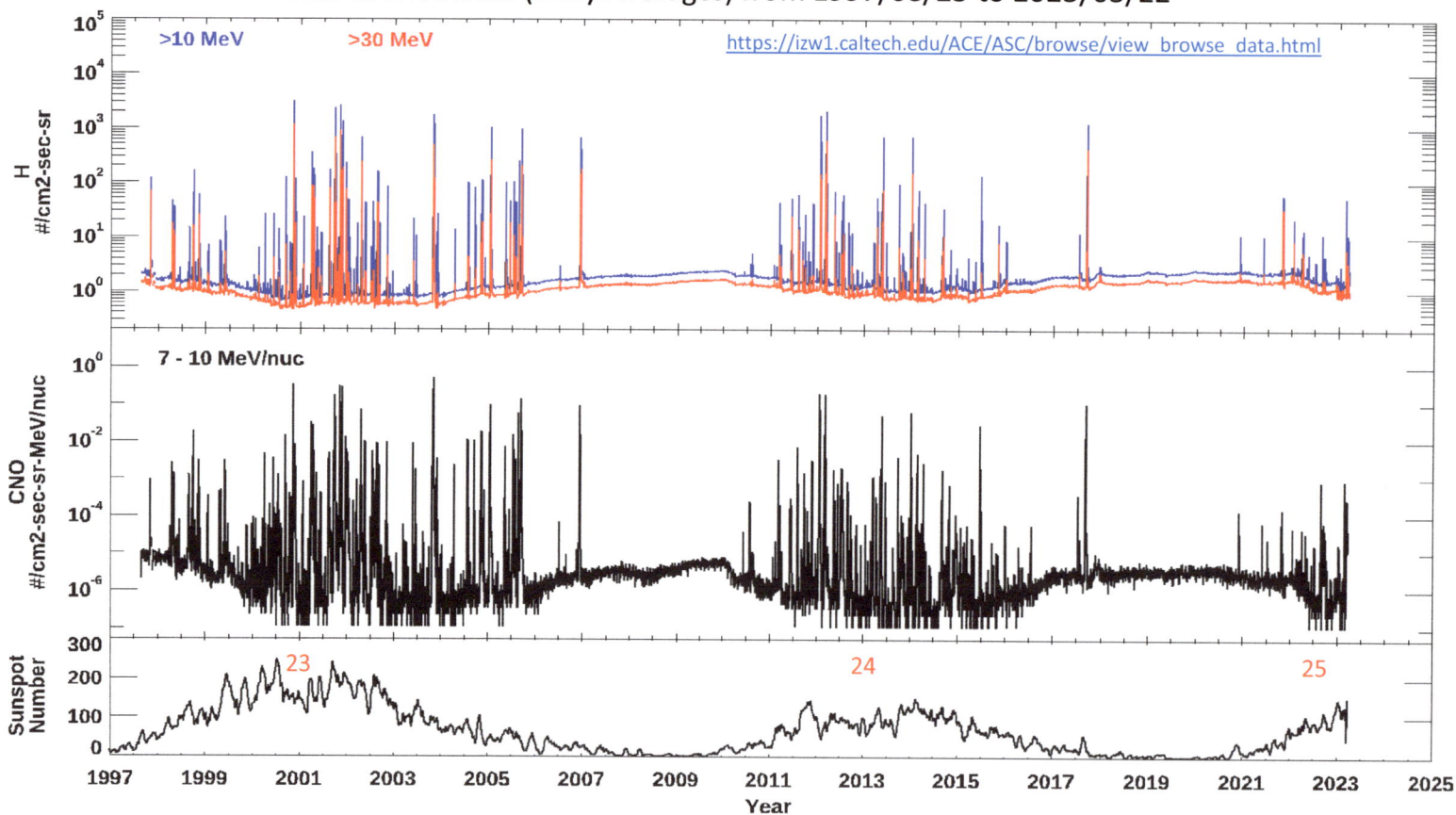

ACE Browse Data (Daily Averages) from 1997/08/25 to 2023/03/22

Geomagnetic Activity: Kp 2019

- Kp index is a measure of the greatest disturbance in the horizontal component of the Earth's magnetic field at mid-latitudes in a 3-hour period: useful index for characterizing the magnitude of geomagnetic storms

- Kp index at solar minimum exhibits a 27-day periodicity due to recurrent geomagnetic storms caused by high speed solar wind flows from coronal holes (solar rotation period is ~27 days)

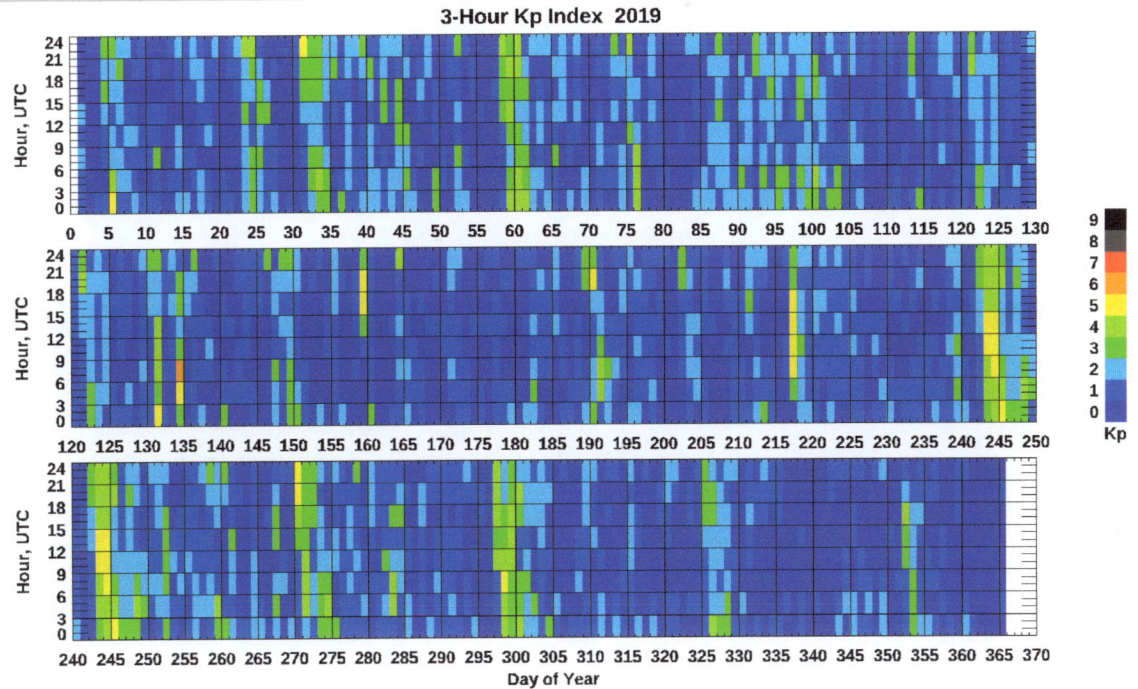

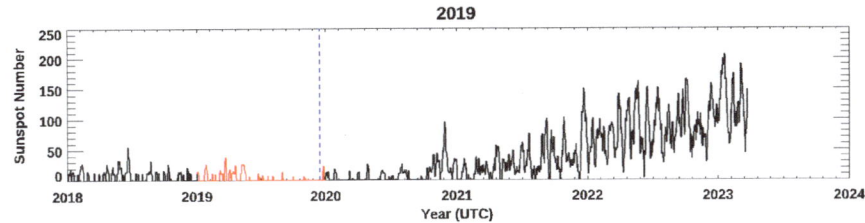

Kp data: GFZ Potsdam
https://kp.gfz-potsdam.de/en/

3-Hour Kp Index 2019

27

- Kp index is a measure of the greatest disturbance in the horizontal component of the Earth's magnetic field at mid-latitudes in a 3-hour period: useful index for characterizing the magnitude of geomagnetic storms

- Kp index at solar minimum exhibits a 27-day periodicity due to recurrent geomagnetic storms caused by high speed solar wind flows from coronal holes (solar rotation period is ~27 days)

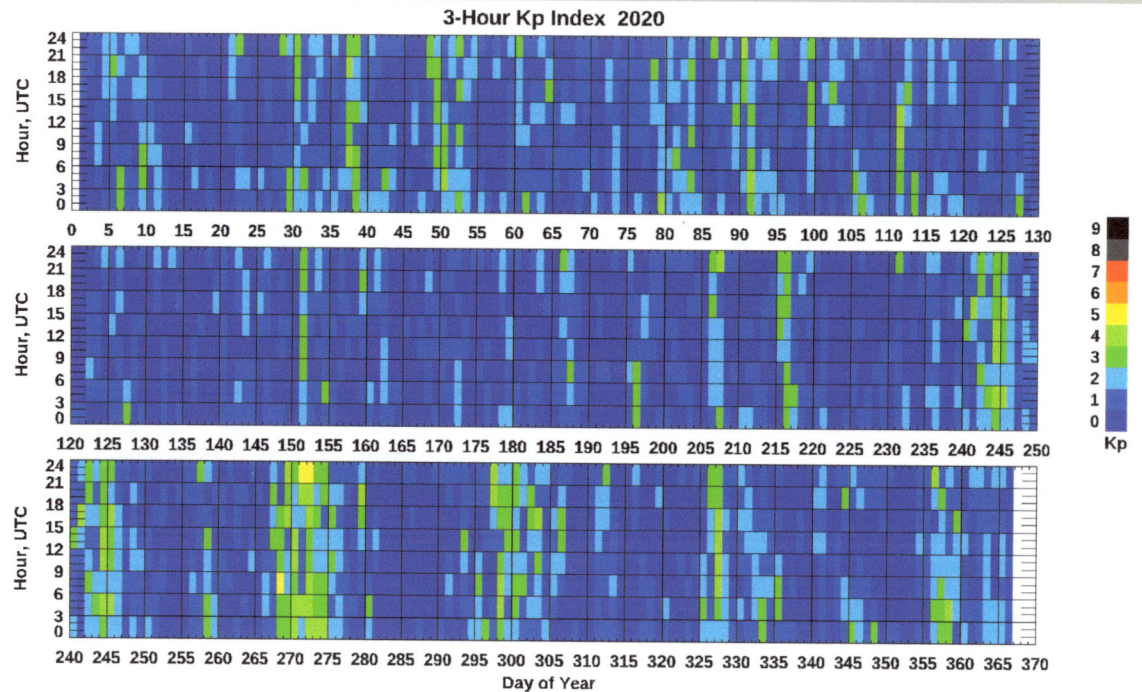

3-Hour Kp Index 2020

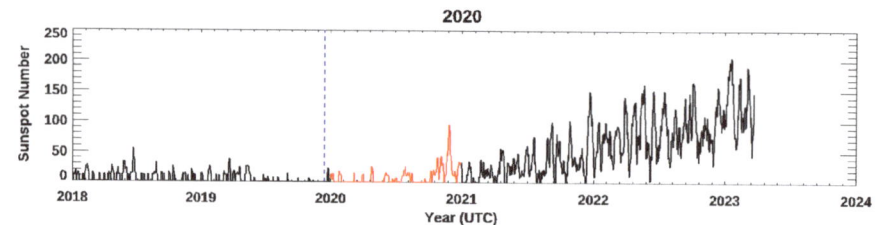

2020

Kp data: GFZ Potsdam
https://kp.gfz-potsdam.de/en/

Geomagnetic Activity: Kp 2021

- Kp index is a measure of the greatest disturbance in the horizontal component of the Earth's magnetic field at mid-latitudes in a 3-hour period: useful index for characterizing the magnitude of geomagnetic storms

- Kp index at solar minimum exhibits a 27-day periodicity due to recurrent geomagnetic storms caused by high speed solar wind flows from coronal holes (solar rotation period is ~27 days)

- CME driven storms increase in number as solar activity increases and coronal hole geometry becomes more complex, washing out the ~27 day periodicity

- Cycle 25 Kp ~ 7 periods first showed up in 2021 for Cycle 25

Kp data: GFZ Potsdam
https://kp.gfz-potsdam.de/en/

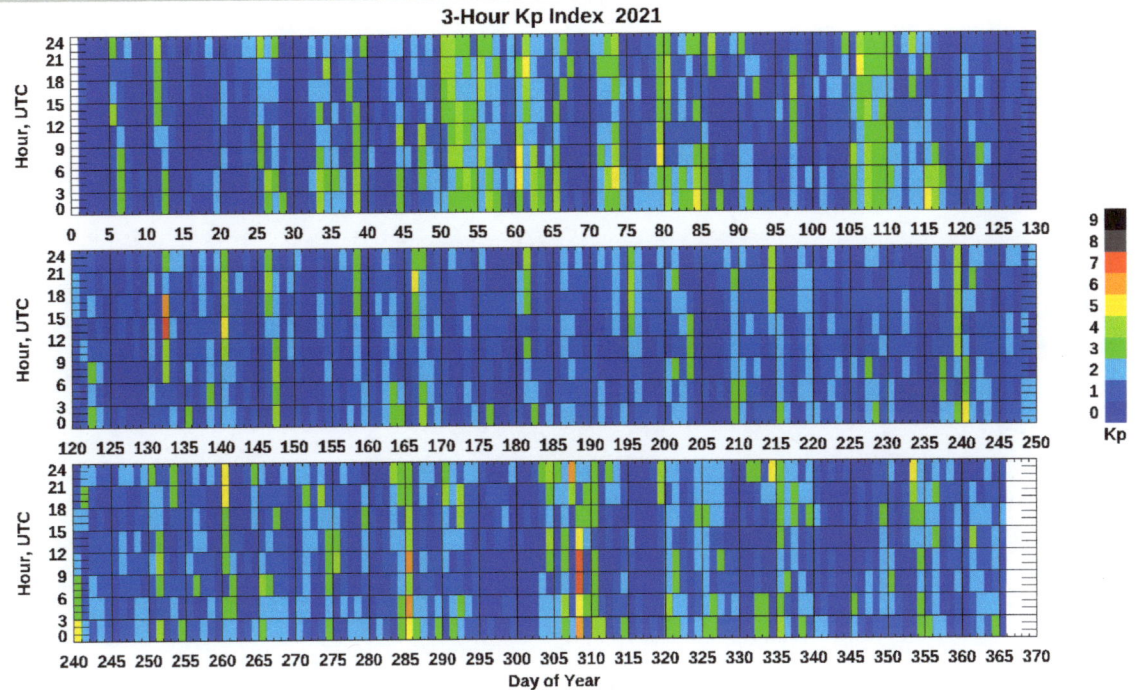

3-Hour Kp Index 2021

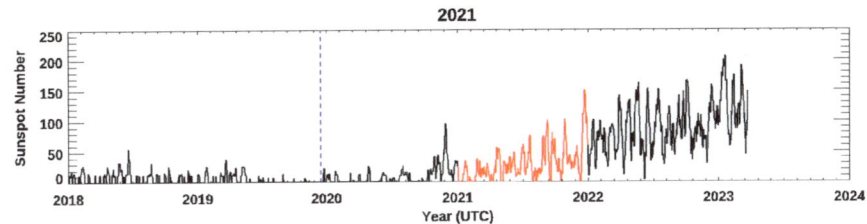

2021

29

Geomagnetic Activity: Kp 2022

- Kp index is a measure of the greatest disturbance in the horizontal component of the Earth's magnetic field at mid-latitudes in a 3-hour period: useful index for characterizing the magnitude of geomagnetic storms

- Kp index at solar minimum exhibits a 27-day periodicity due to recurrent geomagnetic storms caused by high speed solar wind flows from coronal holes (solar rotation period is ~27 days)

- CME driven storms increase in number as solar activity increases and coronal hole geometry becomes more complex, washing out the ~27 day periodicity

- Cycle 25 Kp ~ 7 periods first showed up in 2021 for Cycle 25

Kp data: GFZ Potsdam
https://kp.gfz-potsdam.de/en/

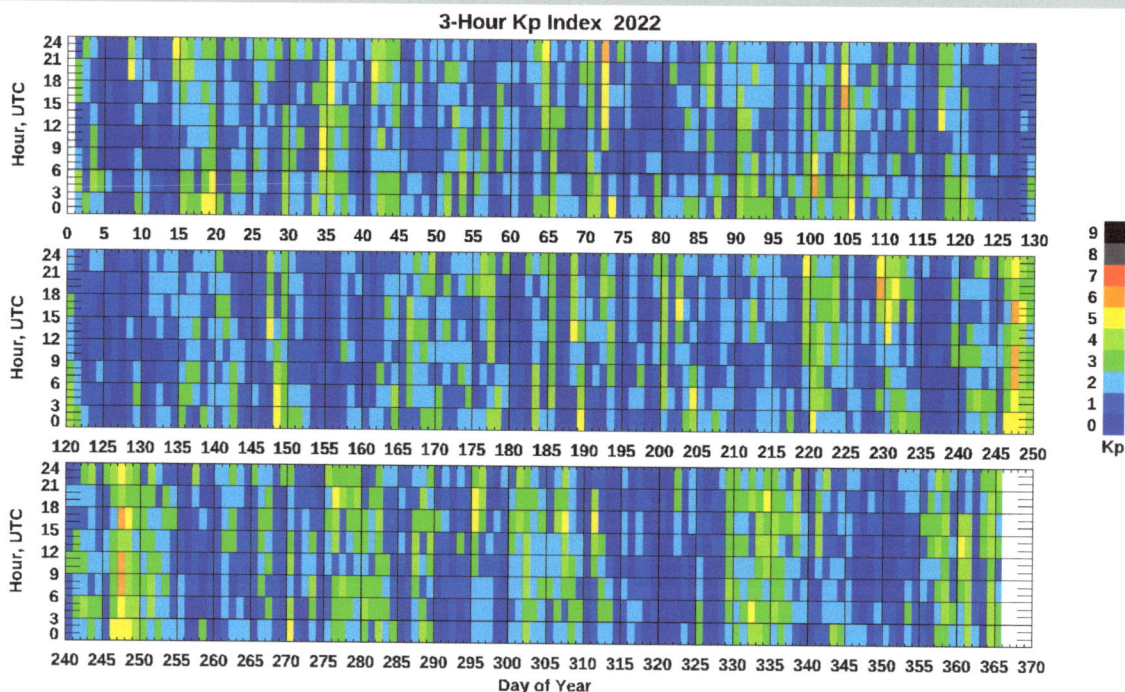

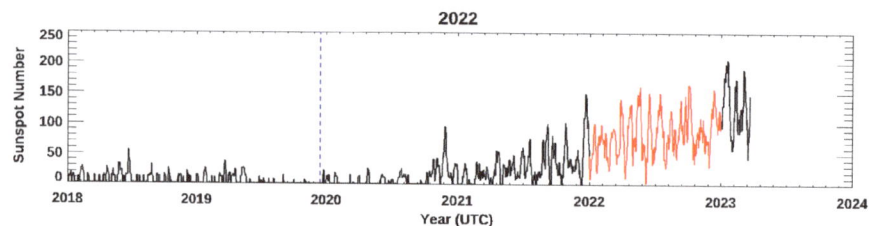

3-Hour Kp Index 2022

30

Geomagnetic Activity: Kp 2023

- Kp index is a measure of the greatest disturbance in the horizontal component of the Earth's magnetic field at mid-latitudes in a 3-hour period: useful index for characterizing the magnitude of geomagnetic storms

- Kp index at solar minimum exhibits a 27-day periodicity due to recurrent geomagnetic storms caused by high speed solar wind flows from coronal holes (solar rotation period is ~27 days)

- CME driven storms increase in number as solar activity increases and coronal hole geometry becomes more complex, washing out the ~27 day periodicity

- Cycle 25 Kp ~ 7 periods first showed up in 2021 for Cycle 25

Kp data: GFZ Potsdam
https://kp.gfz-potsdam.de/en/

3-Hour Kp Index 2023

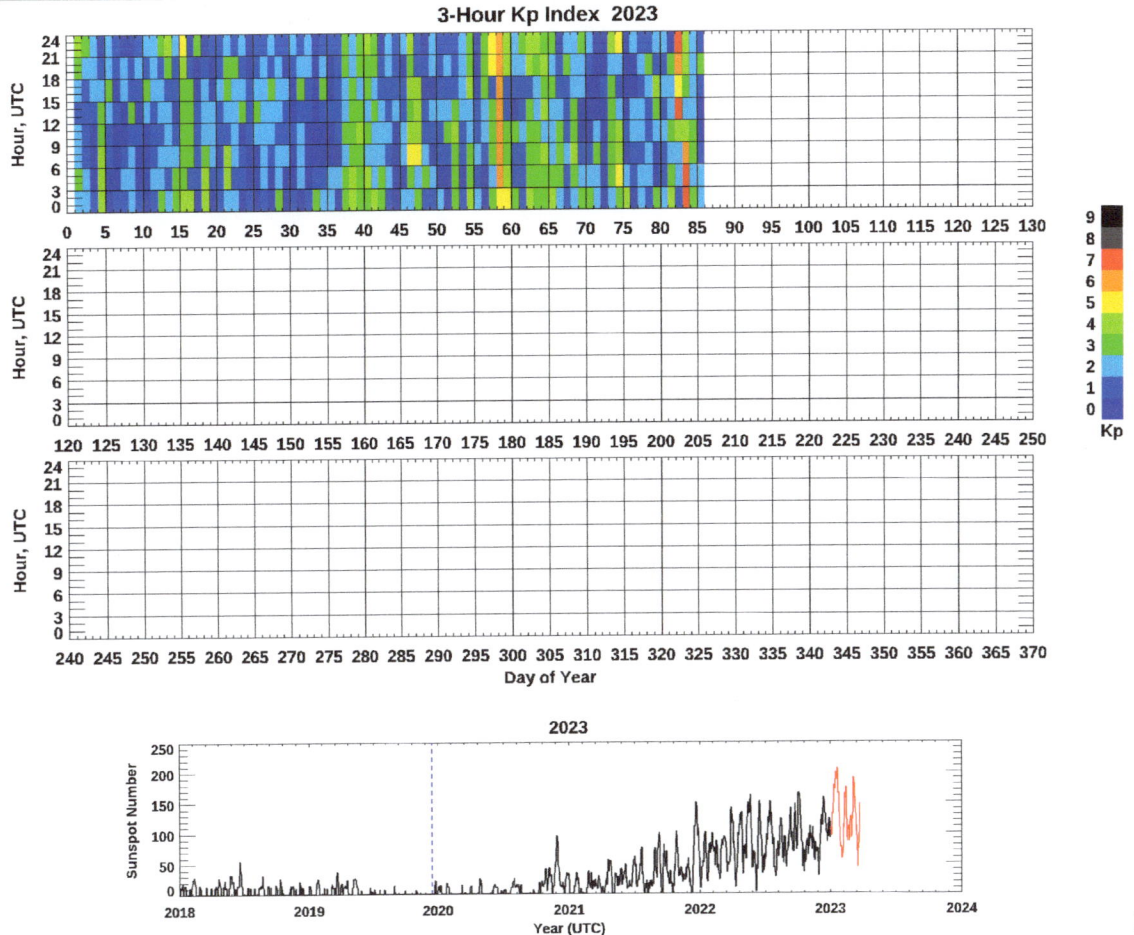

2023

31

Geomagnetic Activity: Kp Statistics

- 3-hr Kp periods with Kp ≥ 6 occurred at the rate of ~30 to 100 periods/year for Solar Cycles 20 – 23

- Cycles 24 and 25 have been less active, only producing Kp ≥ 6 conditions at a rate of ~6 to 30 periods/year

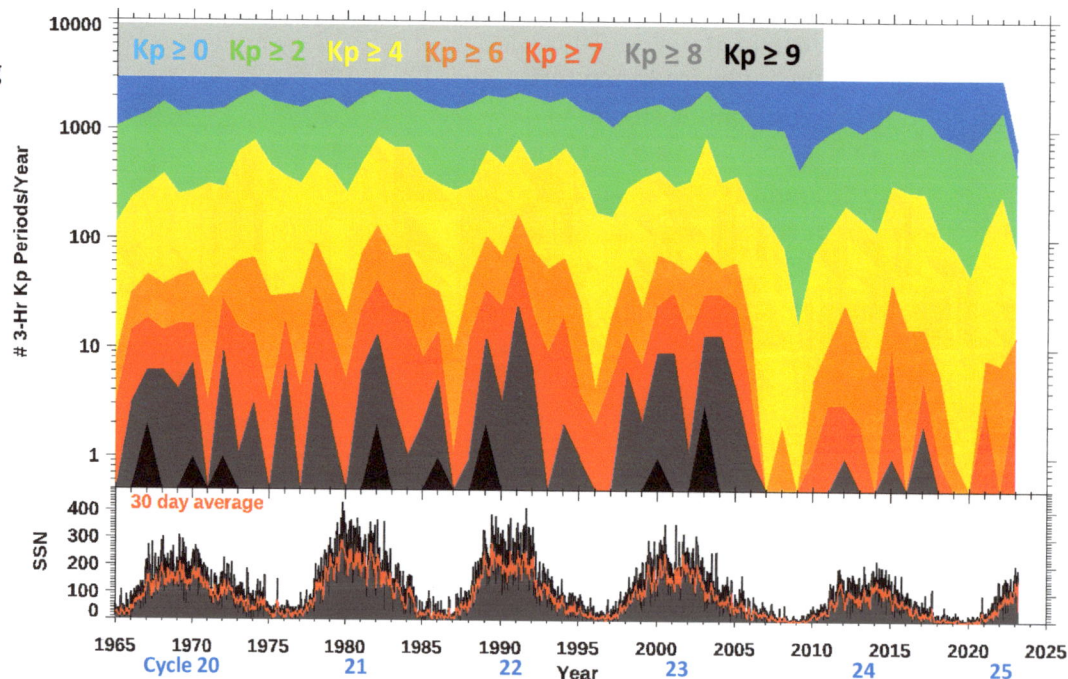

- 3-hr Kp periods with Kp ≥ 6 occurred at the rate of ~30 to 100 periods/year for Solar Cycles 20 – 23

- Cycles 24 and 25 have been less active, only producing Kp ≥ 6 conditions at a rate of ~6 to 30 periods/year

- Cycle 25 appears to be following a similar trend as Cycle 24 with respect to geomagnetic storms:
  - Fewer Kp ≥ 6 periods than previous cycles
  - Very few Kp ≥ 7 periods
  - No Kp ≥ 8 periods to date

- Solar Cycle 25 has produced only seven periods with Kp ≥ 7 (from three storms) to date, no Kp 8 or above:

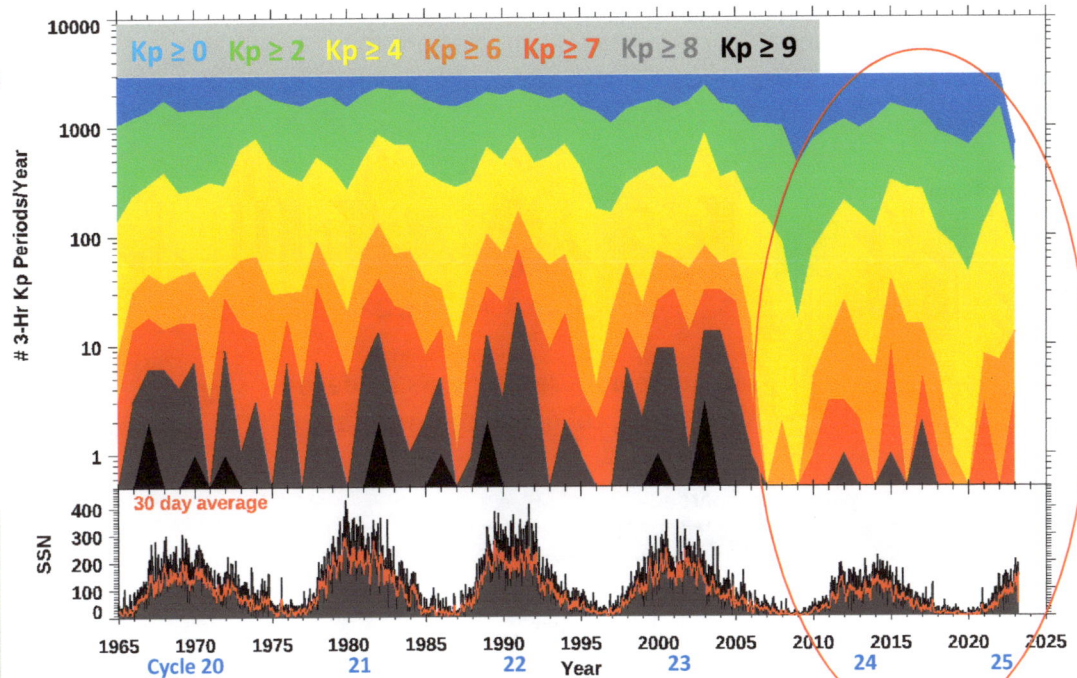

	Number of 3-Hr Kp Periods with Kp ≥ Value								
YYYY	0	1	2	3	4	5	6	7	8	9
2019	2920	1732	770	272	81	21	1	0	0	0
2020	2928	1587	651	217	45	3	0	0	0	0
2021	2920	1853	950	387	123	26	8	3	0	0
2022	2920	2282	1476	697	248	54	7	0	0	0
2023	680	562	392	200	72	23	13	4	0	0

(to 2023/03/26)

Kp data: GFZ Potsdam
https://kp.gfz-potsdam.de/en/

33

Satellite and Orbital Debris Populations

- Active satellite and debris populations continue to grow

- LEO debris environments remain a significant concern for potential collisions and an increasing need for collision avoidance maneuvers

- Large fragmentation debris increases are due to
  1) Fengyun-1C   Chinese ASAT test, 2007
  2) Iridium 33/Cosmos 2251   Accidental collision, 2009
  3) Cosmos 1408   Russian ASAT test, 2021

- Large increase in spacecraft numbers in recent years is due in large part to satellite constellations, cubesats, and smallsats

Space Scoreboard 2023/03/27

Object Type	Objects
Active payloads	7,700
Debris	19,400
Analyst Objects*	19,800
Total	46,800

https://www.space-track.org/

*Analyst objects are variably tracked and in constant flux, Space Track does not provide their catalog and element set data

Cataloged Active and Debris Objects (as of 3 Feb 2023)

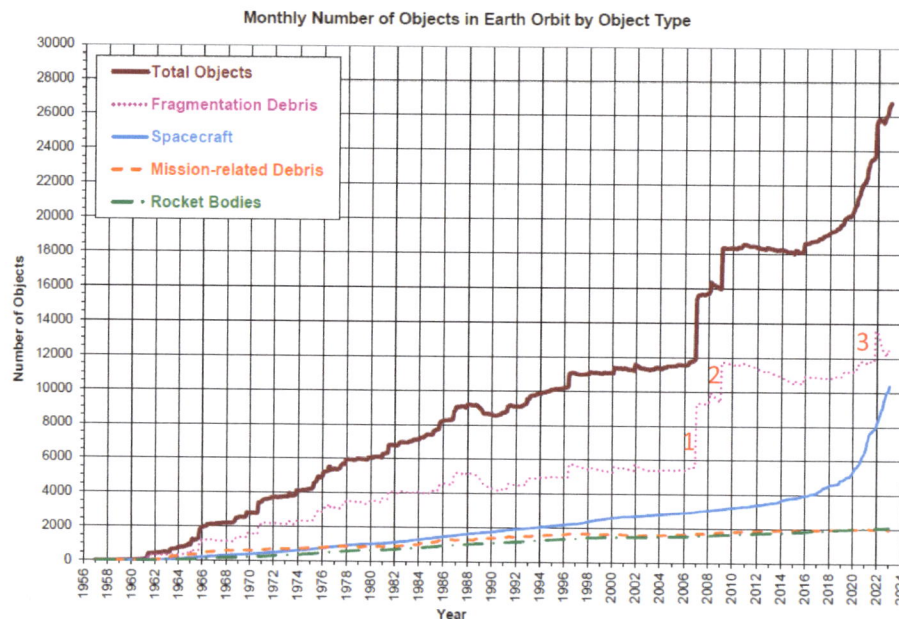

Monthly Number of Objects in Earth Orbit by Object Type

https://orbitaldebris.jsc.nasa.gov/quarterly-news/pdfs/odqnv27i1.pdf

34

On-Orbit Breakups 2020 - 2023

- On-orbit breakups are a significant orbital debris source, producing both short and long term threats to satellites depending on altitude of the breakup
  - Low altitude debris populations reenter quickly
  - High altitude populations will remain in orbit for years to come

- Breakups:
  - Over 250 objects have fragmented in LEO over the last 60 years (LeoLabs, 2022)
  - As of April 2017, 290+ break-ups since 1961 (ESA)

- Causes for breakups include:
  - Residual fuel explosions
  - Power system explosions
  - Accidental collisions with other objects (satellite, debris)
  - Anti-satellite tests
  - Unknown

- Long March 6A breakup, 12 Nov 2022
  - Breakup at 813 x 847 x 98.8° following payload deployment
  - 18th SDS identified 533 fragments from SSN data
  - Hundreds of thousands of fragments as small as 1 mm were likely generated from this breakup.

Sources:
https://blog.leolabs.space/leopulse/quarterly-review-october-2022
https://www.esa.int/ESA_Multimedia/Images/2017/03/Satellite_break-up

On-orbit Breakups: 2020 - 20023

Date	Object	Int'l Designator	Orbit (km x km x deg)	Debris*	Ref
9 Jan 2020	Cosmos 2535	2019-039A	604 x 618 x 97.9°	26	1
12 Feb 2020	SL-14 Tsyklon 3rd Stage	1991-056B	1186 x 1206 x 82.6°	112	1
8 May 2020	SL-23 Zenit Fregat tank	2011—037B	422 x 3606 x 51.5°	325	1
12 Jul 2020	H-2A fairing cover	2018-084C	595 x 643 x 97.9°	87	1,5
27 Aug 2020	Resurs-O1	1994-074A	633 x 660 x 97.9°	72	1
10 Mar 2021	NOAA-17	2002-032A	800 x 817 x 98.62°	102	1,5
18 Mar 2021	Yunhai 1-02	2019-063A	780 x 785 x 98.54°	43	1,5,7
15 Apr 2022	SL-12 (SOZ) Motor	2007-065F	400 x 19068 x 64.8°	19	1
23 Oct 2021	Cosmos 2499	2014-02E	1152 x 1507 x 82.44°	21	1
15 Nov 2021	Cosmos 1408 (ASAT test)	1982-092A	465 x 490 x 82.60°	1600+	1,5
18 Nov 2021	ORBCOMM	1997-084F	758 x 771 x 45°	8	1
26 Nov 2021	Minotaur 4th Stage		567 x 580 x 53.98°	21	1
15 Apr 2022	SL-12 R/B	2007-065F	388 x 19074 x 64.8°	16	5,6
3 Jul 2022	H-2A fairing Cover	2018-084D	579 x 615 x 98°	52	1,5
12 Nov 2022	Long March 6A upper stage	2022-151B	813 x 847 x 98.8°	533	2,5,8
17 Nov 2022	H-2A DEB	2012-025F	609 x 633 x 98.2°	50+	5,8
4 Jan 2023	Cosmos 2499	2014-028E	~1169	85	3,4
10 Feb 2023	SSLV R/B	2023-019D	~350	6	5

*Debris objects in public catalog in addition to parent body   (references in backup)

35

LEO Satellite Populations

- Payload launch traffic into LEO has increased dramatically in recent years with the commercial sector dominating the increase

ESA Space Environment Statistics (*22 December 2022)*

Rocket launches since 1957 (excluding failures)	~ 6,370
Objects placed into Earth orbit	~ 15,070
Objects still in space	~ 9,790
Still functioning	~ 7,200
SSN tracked and cataloged debris	~ 32,000

Debris objects estimated based on statistical models to be in orbit (MASTER-8, future population 2021)

>10 cm	36,500
>1 cm to 10 cm	1,000,000
>1mm to 1 cm	130,000,000

https://sdup.esoc.esa.int/discosweb/statistics/

- SpaceX is a main contributor for the number of new payloads deployed in LEO

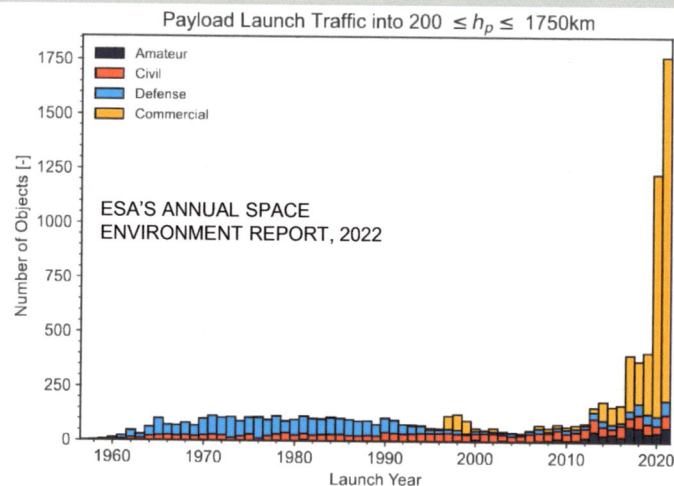

Payload Launch Traffic into 200 $\leq h_p \leq$ 1750km

ESA'S ANNUAL SPACE ENVIRONMENT REPORT, 2022

Legend: Amateur, Civil, Defense, Commercial

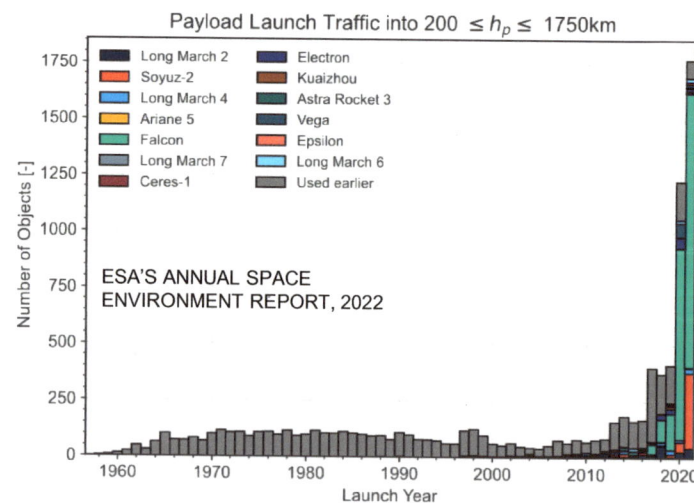

Payload Launch Traffic into 200 $\leq h_p \leq$ 1750km

Legend: Long March 2, Soyuz-2, Long March 4, Ariane 5, Falcon, Long March 7, Ceres-1, Electron, Kuaizhou, Astra Rocket 3, Vega, Epsilon, Long March 6, Used earlier

ESA'S ANNUAL SPACE ENVIRONMENT REPORT, 2022

https://www.sdo.esoc.esa.int/environment_report/Space_Environment_Report_latest.pdf

36

LEO Satellite Populations

- Payload launch traffic into LEO has increased dramatically in recent years with the commercial sector dominating the increase

ESA Space Environment Statistics (*22 December 2022*)	
Rocket launches since 1957 (excluding failures)	~ 6,370
Objects placed into Earth orbit	~ 15,070
Objects still in space	~ 9,790
Still functioning	~ 7,200
SSN tracked and cataloged debris	~ 32,000
Debris objects estimated based on statistical models to be in orbit (MASTER-8, future population 2021)	
>10 cm	36,500
>1 cm to 10 cm	1,000,000
>1mm to 1 cm	130,000,000

https://sdup.esoc.esa.int/discosweb/statistics/

- Traffic characteristics of the LEO operational environment could dramatically change in the coming years if even a fraction of the planned constellations reach their targets for planned spacecraft

Constellation	Alt (km)	In orbit	Active	Planned
Starlink G1 (US)	540-577	3,814	3,165	4,408
Starlink G2A (US)	530-559	239	52	7,500
Starlnk G2 (US)	340-614	---	---	22,488
OneWeb, Phase 1 (UK)	1182-1221	584	426	716
OneWeb, Phase 2 (uk)	1185-1221	---	---	6,372
Kuiper (US)	590-630	---	---	3,232
Guangwan (CH)	508-1145	---	---	12,992
Galaxy Space (CH)	500	---	---	1,000
Hanwha Systems (KR)	500	---	---	2,000
Lynk Global (US)	500	---	---	2,000
Astra (US)	380-700	---	---	13,620
Boeing (US)	670-10,000	---	---	5,841
Telesat (Canada)	1015-1325	---	---	1,969
HVNET (US)	1150	---	---	1140
SpinLaunch (US)	830	---	---	1190
Globalstar3 (Germany)	485-700	---	---	3080
E-SPACE (US)	528-635	---	---	337,323
Total		4,637	3,643	427,171

https://planet4589.org/space/con/conlist.html (17 Mar2023)
https://planet4589.org/space/con/largecon.html (17 Mar 2023)

LEO Satellite Populations

- Payload launch traffic into LEO has increased dramatically in recent years with the commercial sector dominating the increase

ESA Space Environment Statistics (*22 December 2022*)

Rocket launches since 1957 (excluding failures)	~ 6,370
Objects placed into Earth orbit	~ 15,070
Objects still in space	~ 9,790
Still functioning	~ 7,200
SSN tracked and cataloged debris	~ 32,000

Debris objects estimated based on statistical models to be in orbit (MASTER-8, future population 2021)

>10 cm	36,500
>1 cm to 10 cm	1,000,000
>1mm to 1 cm	130,000,000

https://sdup.esoc.esa.int/discosweb/statistics/

- Traffic characteristics of the LEO operational environment could dramatically change in the coming years if even a fraction of the existing and proposed constellations reach their targets for planned spacecraft

Constellation	Alt (km)	In orbit	Active	Planned
Starlink G1 (US)	540-577	3,814	3,165	4,408
Starlink G2A (US)	530-559	239	52	7,500
Starlnk G2 (US)	340-614	---	---	22,488
OneWeb, Phase 1 (UK)	1182-1221	584	426	716
OneWeb, Phase 2 (uk)	1185-1221	---	---	6,372
Kuiper (US)	590-630	---	---	3,232
Guangwan (CH)	508-1145	---	---	12,992
Galaxy Space (CH)	500	---	---	1,000
Hanwha Systems (KR)	500	---	---	2,000
Lynk Global (US)	500	---	---	2,000
Astra (US)	380-700	---	---	13,620
Boei				
Teles				
HVNE				
SpinL				
Glob				
E-SPACE (US)	528-635	---	---	337,323
Total		4,637	3,643	427,171

> In addition, there are many smaller constellations (<1000 spacecraft) planned or in operation representing:
> Total launched: 2244
> Total in operational shells: 956
> Total planned: 4831

https://planet4589.org/space/con/conlist.html (17 Mar2023)
https://planet4589.org/space/con/largecon.html (17 Mar 2023)

- ISS conducted 33 collision avoidance maneuvers since 1999

- Frequency of maneuvers depends on multiple factors including:
  - Solar activity
  - Number of objects crossing ISS orbit
  - Tracking capability of the Space Surveillance Network

- Cosmos 1408 ASAT test generated 1700+ trackable debris objects, ISS has experienced many conjunctions with these objects but only two collision avoidance maneuvers have been executed to date to avoid Cosmos 1408 debris

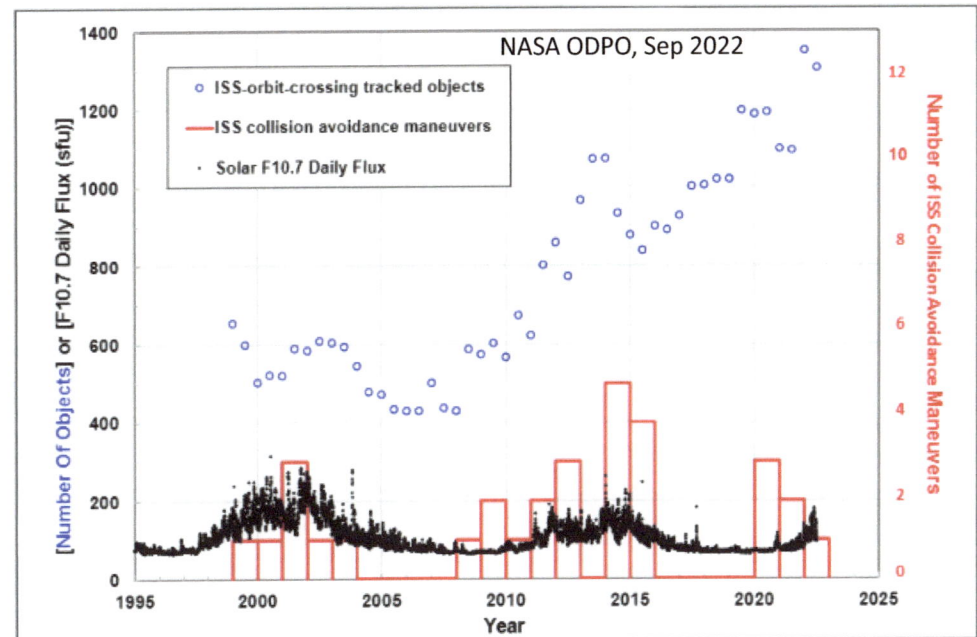

The number of ISS-orbit-crossing objects tracked by the SSN (blue circles), the solar F10.7 daily flux (black dots), and the ISS collision avoidance maneuvers (red histogram) as functions of time.

Sources:
- Orbital Debris Quarterly News, Vol 26, Issue 3, Sept 2022
  Orbital Debris Quarterly News, Vol 26, Issue 4, Dec 2022
  Orbital Debris Quarterly News, Vol 27, Issue 1, Mar 2023
  https://orbitaldebris.jsc.nasa.gov/quarterly-news/
- https://www.space.com/international-space-station-avoid-satellite

39

LRO/Chandrayaan-2 Collision Avoidance Maneuver

- NASA's Lunar Reconnaissance Orbiter (LRO) and ISRO's Chandrayaan-2 (CH2) spacecraft operate in low lunar orbit
  - LRO   20 km x 165 km x polar
  - CH2   100 km x 100 km x polar

- Conjunction on 20 Oct 2021 (DOY 293) at 05:45 UTC with predicted radial separation <100 meters and closest approach of ~3 km

- NASA and ISRO agreed to maneuver CH2 on 18 Oct 2021 to avoid a potential collision. Increased CH2 altitude will prevent future close LRO/CH2 conjunctions

- Collision avoidance maneuvers are a fact of life for low Earth orbit, is this a sign of the future for lunar operations as well as lunar exploration activities increase?

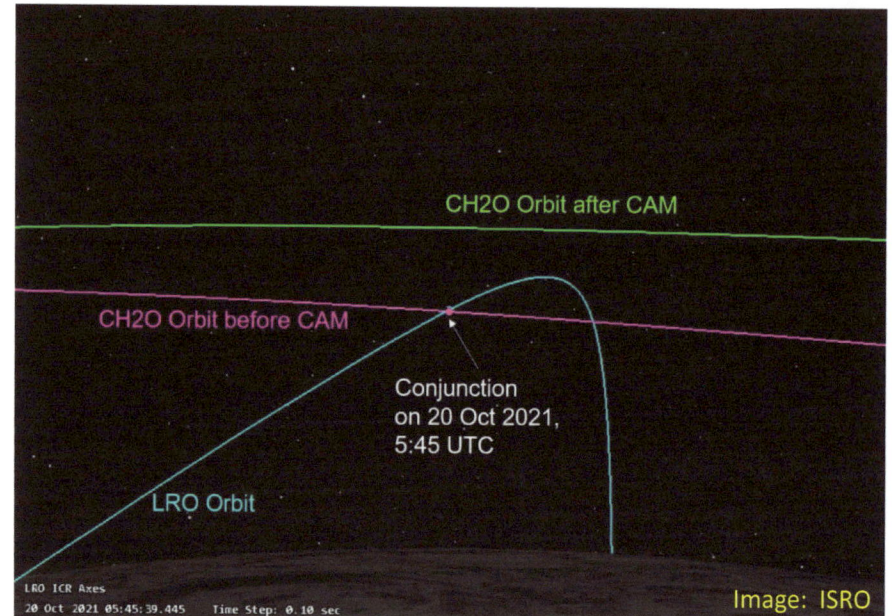

Source:
- https://spacenews.com/indias-chandrayaan-2-maneuvered-to-avoid-close-approach-to-nasas-lunar-reconnaissance-orbiter/
- https://www.cnet.com/google-amp/news/a-spacecraft-made-an-evasive-maneuver-to-avoid-nasas-lunar-orbiter/

Summary

- Space operations continue to be a challenge for both launch vehicles and spacecraft

- Attribution and mitigation of anomalies and failures remains an important concern for space system designers, operators, and users

- The space environment is becoming more hazardous for operations as solar activity increases towards the peak of Solar Cycle 25

- Todays SCAF presentations continue the workshop legacy of reporting case studies relevant to space operations in general and spacecraft anomalies and failures in particular and discussion of methodologies and techniques for evaluating the space environment and its effects on space systems

SPACECRAFT ANOMALIES & FAILURES WORKSHOP
"Creating a Community Solution for Anomaly Attribution"

Co-Sponsored by NRO and NASA
MARCH 29-30, 2023

Locations:
DAY 1 (UNCLASSIFIED): NASA Goddard Space Flight Center
DAY 2 (CLASSIFIED): NRO HQ Westfields

Presentations run from 9:00 AM to 4:00 PM EDT.
Check-In begins at 8:00 AM

AGENDA TOPICS:
- Spacecraft Anomalies, Failures, and Operations

Case Studies
- DART Lessons Learned (Adams)
- Shuttle and Orion Re-entry Temperature Anomalies (Squire)
- JWST Meteoroid Impacts (Menzel)
- Risks for Human Spaceflight Beyond LEO (Vera)
- Space Environments Ops Perspective (Kilzer)

Tools and Resources
- Moon to Mars Space Weather Office (Collado- Vega)
- Long-term Environmental Forecast Tool (O'Brien)
- Models and Applications for Assessing Space Weather from Earth to the Moon (Green)
- NASA Small Spacecraft Anomaly Reporting (Burkhard)

41

Image: NASA

Questions?

Spacecraft	Citation
ELSA-d	1) https://www.space.com/astroscale-suspends-elsa-d-space-debris-cleanup-test 2) https://news.satnews.com/2022/04/07/astroscale-reports-on-their-elsa-d-satellite-servicer-anomalies-during-the-in-space-test-capture/
Starlink	3) https://www.space.com/spacex-starlink-satellite-loss-space-weather-forecast (Group 4-7) 4) https://agupubs.onlinelibrary.wiley.com/doi/full/10.1029/2022SW003193 (Group 4-7) 50) https://www.space.com/james-webb-space-telescope-micrometeoroid-environment (Group 6-1) 51) https://twitter.com/elonmusk/status/1638616130356133888 (Group 6-1, E. Musk Tweet)
SWIFT	5) https://swift.gsfc.nasa.gov/news/2022/safe_mode.html
Aqua	6) https://mcst.gsfc.nasa.gov/news/aqua-safe-mode-event
MAVEN	7) https://www.nasa.gov/feature/goddard/2022/nasa-s-maven-spacecraft-resumes-science-operations-exits-safe-mode 41) https://www.nasa.gov/feature/goddard/2023/maven-status-update
Starliner OFT-2	8) https://www.space.com/nasa-boeing-hail-starliner-launch-success-despite-glitch 9) https://planet4589.org/space/jsr/back/news.806.txt
Vigoride-3	10) https://www.satellitetoday.com/in-space-services/2022/08/02/momentus-identifies-cause-of-vigoride-3-anomaly-deploys-4-more-satellites/ 11) https://space.skyrocket.de/doc_sdat/vigoride.htm 52) https://www.seradata.com/vigoride-3-vr-3-satellite-delivery-spacecraft-has-power-fault-that-meant-only-two-of-its-nine-sat-passengers-got-out/
Cygnus	12) https://www.space.com/space-station-reboost-cygnus-abort (reboost) 13) https://spacenews.com/cygnus-solar-array-fails-to-deploy/ (solar array)
Geotail	14) https://www.nasa.gov/feature/goddard/2023/sun/nasa-s-geotail-mission-operations-come-to-an-end-after-30-years 47) https://blogs.nasa.gov/sunspot/2022/10/17/nasas-geotail-mission-experiences-an-anomaly/
CAPSTONE	15) https://www.space.com/nasa-capstone-cubesat-arrives-moon 16) https://www.space.com/nasa-capstone-moon-probe-anomaly-september-2022
Galaxy-15	17) https://www.space.com/intelsat-loses-control-galaxy-15-satellite-solar-storm 18) https://www.satellitetoday.com/broadcasting/2022/09/01/intelsat-shuts-down-galaxy-15-payload-as-satellite-drifts/ 19) https://spacenews.com/intelsat-working-to-regain-control-of-galaxy-15-satellite/ 55) https://www.seradata.com/galaxy-15-satellite-spirals-out-of-intelsats-control/ 56) https://www.datacenterdynamics.com/en/news/intelsat-turns-off-broadcast-payload-for-wayward-galaxy-15-satellite/

43

References: Spacecraft Anomalies and Failures (public sources)

Spacecraft	Citation
JWST	20) https://www.inverse.com/science/jwst-mrs-anomaly 21) https://scitechdaily.com/webb-space-telescope-glitch-likely-caused-by-galactic-cosmic-ray-niriss-returns-to-full-operations/ 48) https://www.livescience.com/jwst-hit-by-micrometeoroid-space 49) https://www.space.com/james-webb-space-telescope-micrometeoroid-environment
TESS	22) https://www.nasaspaceflight.com/2022/10/tess-safe-mode-incident/
ICON	23) https://blogs.nasa.gov/icon/2022/12/07/icon-mission-out-of-contact/ 24) https://icon.ssl.berkeley.edu/
Orion (Artemis I)	25) https://blogs.nasa.gov/artemis/2022/11/20/artemis-i-flight-day-five-orion-enters-lunar-sphere-of-influence-ahead-of-lunar-flyby/ (Startracker, PCDU) 26) https://blogs.esa.int/orion/2022/11/21/artemis-i-flight-day-1-5/ (Startracker) 28) https://gizmodo.com/nasa-unexpectedly-lost-contact-orion-artemis-1849816978 (Orion comm) 29) https://www.space.com/artemis-1-orion-moon-mission-heat-shield-issue (Orion heat shield)
CubeSats (Artemis I)	27) https://planet4589.org/space/jsr/back/news.813.txt
Lunar Flashlight	30) https://www.jpl.nasa.gov/news/nasas-lunar-flashlight-team-assessing-spacecrafts-propulsion-system
Soyuz	31) https://www.space.com/russian-soyuz-spacecraft-leak-hole-detected 32) https://www.space.com/russian-soyuz-spacecraft-leak-not-geminid-meteor 33) https://spacenews.com/investigation-into-soyuz-leak-continues/ 34) https://www.reuters.com/lifestyle/science/russia-says-leak-soyuz-spacecraft-caused-by-08-millimetre-hole-2022-12-19/ 35) https://spacenews.com/replacement-soyuz-arrives-at-space-station/
Progress	36) https://spaceflightnow.com/2023/02/21/russia-blames-progress-coolant-leak-on-external-influences-as-replacement-soyuz-rolls-to-launch-pad/
Orbiter SN1	37) https://www.space.com/launcher-orbiter-sn1-failure-spacex-transporter-6 38) https://tlpnetwork.com/news/2023/02/launcher-sn1-orbital-vehicle-fails 53) https://www.launcherspace.com/updates/orbiter-sn1-mission-update 54) https://spacenews.com/first-launcher-orbital-transfer-vehicle-fails/

References: Spacecraft Anomalies and Failures (public sources)

Spacecraft	Citation
SWOT	39) https://www.space.com/nasa-swot-water-satellite-instrument-shutdown-february-2023 40) https://blogs.nasa.gov/swot/2023/02/23/engineers-check-swot-science-instrument-during-commissioning-activities/
IBEX	42) https://www.space.com/nasa-ibex-flight-computer-reset-february-2023 43) https://www.space.com/nasa-ibex-flight-computer-normal-march-2023
Falcon-9/EWS RROCI	44) https://space.skyrocket.de/doc_sdat/ews-rroci.htm 45) https://planet4589.org/space/jsr/back/news.816.txt 46) https://www.nasaspaceflight.com/2023/01/spacex-transporter-6/

References: Launch Vehicle Anomalies and Failures (public sources)

Spacecraft	Citation
Multiple LV	1) https://www.space.com/12-biggest-rocket-failures-2022
LV0008	2) https://www.space.com/astra-first-florida-launch-failure-february-2022 3) https://astra.com/news/post-launch-investigation-what-we-found-and-next-steps/
Hyperbola 1	4) https://www.space.com/china-ispace-2nd-launch-failure 5) https://planet4589.org/space/jsr/back/news.806.txt
LV0010	6) https://www.space.com/astra-cancels-rocket-3-production-launch-failures
Small Satellite Launch Vehicle	7) https://thewire.in/space/sslv-suffers-data-loss-at-terminal-stage-isro 8) https://spacenews.com/isro-completes-investigation-into-sslv-launch-failure/
New Shepard	9) https://spacenews.com/blue-origin-says-still-super-early-into-new-shepard-launch-failure-investigation/ 10) https://arstechnica.com/science/2023/01/blue-origin-may-restart-new-shepard-flights-in-april-or-may-or-not/ 42) https://en.wikipedia.org/wiki/New_Shepard 45) https://www.space.com/blue-origin-new-shepard-mishap-engine-nozzle-failure
Alpha	11) https://www.space.com/firefly-alpha-satellite-payloads-reenter-atmosphere 12) https://spacenews.com/firefly-says-alpha-launch-a-success-despite-payload-reentries/
Skylark L	13) https://www.space.com/skyrora-suborbital-rocket-launch-attempt-failure
Epsilon	14) https://www.space.com/japan-epsilon-rocket-launch-failure-october-2022 15) https://www.nasaspaceflight.com/2022/10/epsilon-raise-3/ 16) https://planet4589.org/space/jsr/back/news.811.txt 41) https://spaceflightnow.com/2022/10/18/failure-of-japans-epsilon-rocket-blamed-on-attitude-control-system/
Long March 6A	17) https://www.space.com/12-biggest-rocket-failures-2022 18) https://www.space.com/chinese-rocket-body-breaks-up-after-satellite-launch 39) https://en.wikipedia.org/wiki/Long_March_6A 40) https://en.wikipedia.org/wiki/Long_March_6

References: Launch Vehicle Anomalies and Failures (public sources)

Spacecraft	Citation
Zhuque-2	19) https://spacenews.com/historic-first-launch-of-chinese-private-methane-fueled-rocket-ends-in-failure/ 20) https://interestingengineering.com/innovation/landspace-methane-fueled-rocket-fails 21) https://www.seradata.com/zhuque-2-rocket-fails-to-reach-orbital-velocity-on-maiden-flight/
Vega-C	22) https://www.space.com/arianespace-vega-c-launch-failure-december-2022 23) https://parabolicarc.com/2022/12/20/vega-c-launch-failure-ends-frustrating-year-for-europe/ 36) https://spacenews.com/nozzle-erosion-blamed-for-vega-c-launch-failure/ 38) https://spacenews.com/vega-c-fails-on-second-launch/
LauncherOne	24) https://phys.org/news/2023-01-virgin-orbit-anomaly-satellite-uk.amp 25) https://www.space.com/virgin-orbit-united-kingdom-launch-failure-rocket-part 26) https://www.cbsnews.com/news/virgin-orbit-failure-launcherone-richard-branson/ 37) https://en.wikipedia.org/wiki/LauncherOne
RS1	27) https://www.nasaspaceflight.com/2023/01/abl-rs1-demo-1/ 28) https://www.space.com/abl-space-systems-debut-launch-failure 29) https://planet4589.org/space/jsr/back/news.816.txt 30) https://spaceflightnow.com/2023/01/11/first-launch-by-abl-space-systems-fails-shortly-after-liftoff/
H3	34) https://www.space.com/japan-h3-rocket-fails-1st-test-flight 35) https://global.jaxa.jp/press/2023/03/20230305-1_e.html
Terran-1	43) https://spaceflightnow.com/2023/03/22/relativity-space-terran-1-glhf-2/ 44) https://www.space.com/relativity-space-terran-1-test-launch-failure

References: On-Orbit Breakups (public sources)

Citation
1) https://orbitaldebris.jsc.nasa.gov/quarterly-news/
2) https://www.space.com/12-biggest-rocket-failures-2022 https://www.space.com/chinese-rocket-body-breaks-up-after-satellite-launch
3) https://www.space.com/russian-satellite-kosmos-2499-breakup-earth-orbit
4) https://twitter.com/18thsds?lang=en
5) https://www.space-track.org/
6) https://twitter.com/planet4589/status/1521541466153263105
7) https://spacenews.com/breakup-of-chinas-yunhai-1-02-satellite-linked-to-space-debris-collision/
8) https://orbitaldebris.jsc.nasa.gov/quarterly-news/pdfs/odqnv27i1.pdf

ACE	Advanced Composition Explorer (s/c)	JAXA	Japan Aerospace Exploration Agency	SSN	sunspot number
Ap	daily planetary magnetic index	JWST	James Webb Space Telescope (s/c)	SWIFT	Neil Gehrels Swift Observatory (s/c)
ap	3-hour planetary magnetic index	km	kilometers	SWOT	Surface Water and Ocean Topography (s/c)
ASAT	anti-satellite	Kp	planetary K index	SWx	space weather
CAPSTONE	Cislunar Autonomous Positioning System Technology Operations and Navigation Experiment (s/c)	LaRC	Langley Research Center	TESS	Transiting Exoplanet Survey Satellite (s/c)
		LASCO	Large Angle and Spectrometric Coronagraph Experiment	US	United States
				USSF	US Space Force
CCMC	Community Coordinated Modeling Center	LEO	low Earth orbit	3-D	three dimensional
CDAWeb	Coordinated Data Analysis Web	LM	Long March (LV)		
CH2	Chandrayaan-2 (s/c)	LRO	Lunar Reconnaissance Orbiter (s/c)		
CME	coronal mass ejection	LV	launch vehicle		
CNO	carbon, nitrogen, oxygen	L2	second Lagrange point		
CSA	Canadian Space Agency	MAVEN	Mars Atmosphere and Volatile Evolution (s/c)		
DEB	debris	MeV	million electron volts		
DONKI	Database of Notifications, Knowledge, Information	mm	millimeter		
Dst	disturbance storm-time index	M/OD	meteoroids and orbital debris		
EOL	end of life	MS	Soyuz model		
ESLA-d	End-of-Life Services by Astroscale-demonstration (s/c)	MSFC	Marshall Space Flight Center		
		M2M	Moon to Mars Space Weather Analysis Office		
ESA	European Space Agency	NASA	National Aeronautics and Space Administration		
EUV	extreme ultraviolet	NG	Northrop Grumman		
EVA	extravehicular activity	NOAA	National Oceanic and Atmospheric Administration		
EWS RROCI	Electro Optical/Infrared Weather System Rapid Revisit Optical Cloud Imager (s/c)	NRO	National Reconnaissance Office		
		nuc	nucleon		
F9	Falcon 9	OFT	Orbital Flight Test		
F107	10.7 cm radio flux (solar EUV proxy)	R/B	rocket body		
GCR	galactic cosmic rays	Re	Earth radius		
GFZ	*GeoForschungsZentrum* (Geo-research Centre)	s/c	spacecraft		
GSFC	Goddard Space Flight Center	18th SDS	18th Space Defense Squadron		
H	hydrogen	SOHO	Solar and Heliospheric Observatory (s/c)		
IAI	Integrity Applications Incorporated	SPE	solar particle event		
IBEX	Interstellar Boundary Explorer (s/c)	SCAF	Spacecraft Anomalies and Failures		
ICON	Ionospheric Connections Explorer (s/c)	SL	Russian rocket body		
ISRO	Indian Space Research Organization	SSLV	Small Satellite Launch Vehicle (LV))		
ISS	International Space Station	SSN	Space Surveillance Network		

49

www.ingramcontent.com/pod-product-compliance
Lightning Source LLC
Chambersburg PA
CBHW041448200326
41518CB00004B/189